Berry Crop Production and Protection

Berry Crop Production and Protection

Special Issue Editor

Samir C. Debnath

MDPI • Basel • Beijing • Wuhan • Barcelona • Belgrade

MDPI

Special Issue Editor
Samir C. Debnath
Agriculture and Agri-Food Canada,
St. John's, NL, Canada

Editorial Office
MDPI
St. Alban-Anlage 66
4052 Basel, Switzerland

This is a reprint of articles from the Special Issue published online in the open access journal *Agronomy* (ISSN 2073-4395) from 2018 to 2019 (available at: https://www.mdpi.com/journal/agronomy/special_issues/berry_production_protection)

For citation purposes, cite each article independently as indicated on the article page online and as indicated below:

LastName, A.A.; LastName, B.B.; LastName, C.C. Article Title. *Journal Name* **Year**, *Article Number*, Page Range.

ISBN 978-3-03921-094-7 (Pbk)
ISBN 978-3-03921-095-4 (PDF)

Contents

About the Special Issue Editor

Samir C. Debnath, Dr. is a Research Scientist at the St. John's Research and Development Centre of Agriculture and Agri-Food Canada (AAFC) in St. John's, Newfoundland and Labrador (NL), Canada and an Adjunct Professor of Biology at the Memorial University of Newfoundland, St. John's, NL, Canada.

Dr. Debnath has authored and co-authored more than 130 peer-reviewed journal articles including review papers and book chapters, and numerous other publications in plant propagation, biotechnology, biodiversity, and breeding. He has been a keynote speaker and an invited speaker at a number of international and national conferences and meetings, is an active member of some national and international professional associations, and was the President of the Newfoundland and Labrador Institute of Agrologists (P.Ag.) and the Canadian Society for Horticultural Science. Dr. Debnath was an Editor-in-Chief of *Scientia Horticulturae*, Special Issue Editor of *Agronomy*, Associate Editor of the *Canadian Journal for Horticultural Science* and the *Journal of Horticultural Science and Biotechnology*, and was also the Country (Canada) Representative and Council Member of the International Society for Horticultural Science.

Dr. Debnath's research concerns biotechnology-based value-added small fruit and medicinal plant improvement. Much of his current work focuses on wild germplasm, biodiversity, propagation, and improvement of small fruit (berry) crops including blueberry, cranberry, lingonberry, strawberry, raspberry, and cloudberry, and medicinal plants (roseroot) using in vitro (bioreactor micropropagation, in vitro selection, protoplast fusion) and molecular techniques (gene editing, clonal fidelity, genetic diversity, marker-assisted selection) combined with conventional methods.

Preface to "Berry Crop Production and Protection"

Berry crops include, but are not limited to, the genera: *Fragaria* (strawberry, Rosaceae), *Ribes* (currant and gooseberry, Grossulariaceae), *Rubus* (brambles: raspberry and blackberry; Rosaceae), *Vaccinium* (blueberry, cranberry, and lingonberry; Ericaceae) and *Vitis* (grapes, Vitaceae). They possess economically important variously colored, soft-fleshed, small fruits that are grown all over the world. These fruits are consumed fresh or frozen, and are also processed as functional food supplements in industrial products. The significant role of these fruits in maintaining human health has dramatically increased their popularity and production across the world. This Special Issue Book covers berry crops in nine chapters, including one review paper. Various areas of production systems, propagation, plant and soil nutrition, health benefits, marketing and economics, and other related areas have been covered. The aim was to bring together a collection of valuable articles that would serve as a foundation of innovative ideas for the production and protection of health-promoting berry crops in a changing environment.

Samir C. Debnath
Special Issue Editor

![agronomy logo] *agronomy*

MDPI

Article

Application of Nano-Silicon Dioxide Improves Salt Stress Tolerance in Strawberry Plants

Saber Avestan [1,*], Mahmood Ghasemnezhad [1], Masoud Esfahani [1] and Caitlin S. Byrt [2]

[1]   Department of Horticultural Science, University of Guilan, Rasht 4199613776, Iran;
      Ghasemnezhad@Guilan.ac.ir (M.G.); Esfahani@Guilan.ac.ir (M.E.)
[2]   School of Agriculture, Food and Wine, University of Adelaide, Urrbrae 5064, South Australia, Australia;
      caitlin.byrt@adelaide.edu.au
*   Correspondence: avestansaber@phd.guilan.ac.ir; Tel.: +98-1333367343

Received: 18 April 2019; Accepted: 8 May 2019; Published: 17 May 2019

Abstract: Silicon application can improve productivity outcomes for salt stressed plants. Here, we describe how strawberry plants respond to treatments including various combinations of salt stress and nano-silicon dioxide, and assess whether nano-silicon dioxide improves strawberry plant tolerance to salt stress. Strawberry plants were treated with salt (0, 25 or 50 mM NaCl), and the nano-silicon dioxide treatments were applied to the strawberry plants before (0, 50 and 100 mg L^{-1}) or after (0 and 50 mg L^{-1}) flowering. The salt stress treatments reduced plant biomass, chlorophyll content, and leaf relative water content (RWC) as expected. Relative to control (no NaCl) plants the salt treated plants had 10% lower membrane stability index (MSI), 81% greater proline content, and 54% greater cuticular transpiration; as well as increased canopy temperature and changes in the structure of the epicuticular wax layer. The plants treated with nano-silicon dioxide were better able to maintain epicuticular wax structure, chlorophyll content, and carotenoid content and accumulated less proline relative to plants treated only with salt and no nano-silicon dioxide. Analysis of scanning electron microscopic (SEM) images revealed that the salt treatments resulted in changes in epicuticular wax type and thickness, and that the application of nano-silicon dioxide suppressed the adverse effects of salinity on the epicuticular wax layer. Nano-silicon dioxide treated salt stressed plants had increased irregular (smoother) crystal wax deposits in their epicuticular layer. Together these observations indicate that application of nano-silicon dioxide can limit the adverse anatomical and biochemical changes related to salt stress impacts on strawberry plants and that this is, in part, associated with epicuticular wax deposition.

Keywords: abiotic stress; epicuticular wax; nanoparticle; silicon

---

1. Introduction

Plants routinely experience adverse environmental conditions during their growth and development. For example, conditions such as drought, salinity, and cold stress frequently have adverse effects on plant growth and metabolism [1,2]. Salt or salinity stress may have a negative effect on the growth, development, and even survival of the plant by imposing osmotic stress along with causing ion and nutritional imbalances. The application of additional nutrients, such as calcium, can be considered as one strategy to reduce the effects of the ionic imbalance and plant nutritional deficiencies that occur in saline soils, and application of silicon can also improve outcomes for plants growing in saline soils [3]. Strawberries are relatively sensitive to salinity, and salinity can cause leaf burns, necrosis, nutritional imbalance, or specific ionic toxicity (due to sodium and chloride accumulation); this decreases the quality and yield of fruit, and increases the probability of plant mortality [4]. Exploring salt stress responses in strawberry is also of interest because strawberry is a model for the study of the Rosaceae family [5].

Silicon is not classed as an essential nutrient, but it is involved in a number of metabolic pathways that increase the tolerance of plants to environmental stress, such as drought and salinity stress [6–8]. Application of silicon is associated with increased resistance to water loss and improvement in plant water status in saline conditions, relative to control plants [9,10]. Silicon deposits have been observed in epidermal cell walls and this deposition is associated with limiting water loss from the cuticle and excessive transpiration [11]. Previous studies have linked silicon application, in the context of salinity, with enhanced photosynthesis, increased vegetative growth and dry matter production, reduced shoot sodium, and chloride accumulation and increased potassium accumulation and reduced root-to-shoot boron transport [12–14]; therefore, further research is needed towards determining the complement of reasons why silicon application benefits plants [6].

One way in which silicon may be applied to plants is in the form of nanoparticles. Application of silicon nanoparticles is reported to be an effective alternative to adding silicon as part of conventional mineral fertilizers [15]. For example, Prasad et al. [16] reported that zinc nanoparticles improved seed germination, plant growth, flowering, chlorophyll content and yield of peanut (*Arachis hypogaea* L.) compared to zinc sulfate treatments. In addition, it has been suggested that silica oxide nanoparticles can increase cell wall thickness, which can inhibit the penetration of fungi, bacteria and nematodes, and increase resistance to disease [16]. Silicon accumulation in plants is also linked to epicuticular wax accumulation. For example, in cucumber (*Cucumis sativus* L., cv. Corona) changes in the fruit trichome morphology occurred in response to silicon application and the silica accumulation was restricted to the trichomes, primarily in the epicuticular wax [17]. Epicuticular wax accumulation is linked to plant water use efficiency and the regulation of the amount of moisture evaporation through the leaf [18,19]. Therefore, increasing the amount of epicuticular wax may be a type of adaptation to environmental stresses [20]. As wax deposition plays a protective role against water loss through the cuticle, increasing wax content is classified as a dehydration avoidance mechanism [19]. The aim of this study was to investigate whether application of nano-silicon dioxide suppressed the adverse effects of salt stress on strawberry (*Fragaria* × *anansa* Duch.) plant growth and development, and to study how nano-silicon dioxide application might influence changes in anatomy and biochemistry previously linked with salt stress and silicon treatments.

2. Materials and Methods

2.1. Growth Conditions and Treatments

The experiment was conducted under greenhouse conditions at University of Guilan, Rasht, Iran. Strawberry (*Fragaria* × *anansa*) plants 'cv; Camarosa' with 11 mm crown diameters were obtained from a commercial nursery in Kurdistan province, Iran. Nano-particles of silicon dioxide were obtained from Sigma-Aldrich (Lot 637238). Nano-silicon dioxide characteristics were: 99.5% purity and 10–20 nm particle size, and particles were applied as a suspension phase (suspended in nutrient solution) relative to control (no $nSiO_2$) treatments of only nutrient solution. The strawberry plants (*Fragaria* × *anansa*, 'cv; Camarosa') were grown in the following conditions: 12 h photoperiod, 25 ± 10 °C temperature, 70 ± 10% relative humidity. Plants with 11 mm crown diameters (approximately 40 days old) which had received two weeks chilling requirement were transferred to a greenhouse and planted into 4 L containers filled with coco-peat and perlite (2/1, *v/v*). The plants were fertilized with modified Hoagland's solution with or without nano-silicon dioxide. Two different nutrient solutions were used in this experiment to meet plant nutritional needs during vegetative growth and at flowering. Before the start of flowering; the nutrient solution contained elemental concentrations as follows: 150 mg L^{-1} N, 54 mg L^{-1} P, 262 mg L^{-1} K, 110 mg L^{-1} Ca, 34 mg L^{-1} Mg, 50 mg L^{-1} S, 5 mg L^{-1} Fe, 0.5 mg L^{-1} Mn, 0.5 mg L^{-1} Zn, 0.50 mg L^{-1} B, 0.05 mg L^{-1} Cu and 0.05 mg L^{-1} Mo. During flowering, the nutrient solution contained 142 mg L^{-1} N, 59 mg L^{-1} P, 227 mg L^{-1} K, 110 mg L^{-1} Ca, 39 mg L^{-1} Mg, 56 mg L^{-1} S, 5 mg L^{-1} Fe, 0.5 mg L^{-1} Mn, 0. 5 mg L^{-1} Zn, 0.50 mg L^{-1} B, 0.05 mg L^{-1}

Cu and 0.05 mg L^{-1} Mo. The pH of nutrient solution was adjusted to 6. The nano-silicon dioxide (0, 50, 100 mg L^{-1}) was incorporated into the Hoagland's solution nutrients.

Salt stress treatments were imposed by dissolving NaCl (to achieve 0, 25 and 50 mM concentrations) into the nutrient solution which was used to water the plants (see Table 1). The plants were exposed to salt stress two weeks after planting. In order to prevent salt stress shock, salt concentrations were increased gradually during the first two weeks of the salt stress and after this period saline solution was applied every four days. In addition, the containers were irrigated with 600 mL water for leaching salt every two weeks during salinity treatment.

Table 1. Combinations of nano-silicon dioxide and salinity stress treatments tested.

Salinity (mM)	nSiO$_2$ mg L^{-1} before BBCH: 61	nSiO$_2$ mg L^{-1} after BBCH: 61	Treatments	
0 mM (Control—no NaCl)	0	0 (Control—no NaCl, no SiO$_2$)	S$_1$	0 mM NaCl + 0 mg L^{-1} nSiO$_2$
	50	50	S$_2$	0 mM NaCl + 0.50 mg L^{-1} SiO$_2$
		0	S$_3$	0 mM NaCl + 50. 0 mg L^{-1} SiO$_2$
		50	S$_4$	0 mM NaCl + 50.50 mg L^{-1} SiO$_2$
	100	0	S$_5$	0 mM NaCl + 100.0 mg L^{-1} SiO$_2$
		50	S$_6$	0 mM NaCl + 100.50 mg L^{-1} SiO$_2$
25 mM	0	0 (Control—no SiO$_2$)	S$_1$	25 mM NaCl + 0 mg L^{-1} nSio$_2$
	50	50	S$_2$	25 mM NaCl + 0.50 mg L^{-1} SiO$_2$
		0	S$_3$	25 mM NaCl + 50.0 mg L^{-1} SiO$_2$
		50	S$_4$	25 mM NaCl + 50.50 mg L^{-1} SiO$_2$
	100	0	S$_5$	25 mM NaCl + 100.0 mg L^{-1} SiO$_2$
		50	S$_6$	25 mM NaCl + 100.50 mg L^{-1} SiO$_2$
50 mM	0	0 (Control—no SiO$_2$)	S$_1$	50 mM NaCl + 0 mg L^{-1} nSio$_2$
	50	50	S$_2$	50 mM NaCl + 0.50 mg L^{-1} SiO$_2$
		0	S$_3$	50 mM NaCl + 50.0 mg L^{-1} SiO$_2$
		50	S$_4$	50 mM NaCl + 50.50 mg L^{-1} SiO$_2$
	100	0	S$_5$	50 mM NaCl+ 100.0 mg L^{-1} SiO$_2$
		50	S$_6$	50 mM NaCl+ 100.50 mg L^{-1} SiO$_2$

BBCH: Biologische Bundesanstalt, Bundessortenamt und CHemische Industrie. S$_1$ = control (no nSiO$_2$ application before or after BBCH: 61). S$_2$ = 50 mg L^{-1} nSiO$_2$ just after BBCH: 61. S$_3$ = 50 mg L^{-1} nSiO$_2$ before BBCH: 61. S$_4$ = 50 mg L^{-1} nSiO$_2$ throughout all growth and development stages. S$_5$ = 100 mg L^{-1} nSiO$_2$ before BBCH: 61. S$_6$ = 100 mg L^{-1} nSiO$_2$ before BBCH: 61 and 50 mg L^{-1} after BBCH: 61.

The plants were treated with the following concentrations of nano-silicon dioxide: 0, 50, 100 mg L^{-1} after planting until the beginning of flowering: when about 10% of flowers had started to open (BBCH (Biologische Bundesanstalt, Bundessortenamt und CHemische Industrie): 61) or were at vegetative stages (phenological growth stages and BBCH-identification keys of strawberry (*Fragaria* × *ananassa* Duch.). Thereafter, the plants were treated continuously during the reproductive stage (BBCH: 61–92) with treatments of 0, or 50 mg L^{-1} nano-silicon dioxide concentrations; the nSiO$_2$ treatments were divided into six groups (Table 1):

2.2. Phenotypic Measurements

The fresh weight of shoots and root were measured at the end of the experiment, and harvested samples were immediately dried in an oven at 70 °C for 48 h, and subsequently, the dry weight was determined.

Relative water contents (RWC) of leaves were determined according to Abdi et al. [21] and calculated using the following Equation:

$$RWC = (FW - DW)/(TW - DW) \times 100 \qquad (1)$$

where FW (fresh weight) of the leaves was measured immediately after picking and DW (dry weight) was measured after drying the leaves in an oven at 70 °C for 24 h or until constant weight was achieved; the leaf weight at full turgor was TW, measured after floating the leaves for 4 h in distilled water at room temperature in the dark [21]; three biological replicates per treatment were included.

Relative water protection (RWP): three comparable leaves were randomly selected from three biological plant replicates were weighed to determine fresh weight (FW) and thereafter allowed to wilt at 25 °C for 8 h then weighed (Withering weight, WW). The samples were oven-dried at 70 °C for 72 h and reweighed (Dry weight, DW). Finally, RWP was calculated following [22]:

$$RWP = ((WW - DW)/(FW - DW)) \qquad (2)$$

Relative water loss (RWL): three comparable leaves were removed from each plant (three biological replications per treatment) and immediately weighed (W1). The leaves were allowed to wilt at 25 °C and weighed over 2, 5 and 8 h (W2, W3, and W4). The samples were oven-dried at 70 °C for 72 h and reweighed (Wd). RWL was calculated by the following formula [23].

$$RWL = ((W1 - W2) + (W2 - W4))/((3 \times WD \ (T1 - T2)). \qquad (3)$$

Membrane stability index (MSI) was measured following Sairam [24]. The leaf sections, 5 cm^2 were put in 10 mL of double-distilled water. One set was kept for 30 min at 40 °C and its electrical conductivity recorded using a conductivity meter (C1), while the second set was kept for 10 min in a boiling water bath (100 °C) and subsequently measurements of conductivity were taken (C2). The electrolyte leakage or membrane stability index were calculated following [24]:

$$MSI = (1 - (C1/C2)) \times 100 \qquad (4)$$

Cuticle transpiration (CT): The cuticle transpiration was calculated using the following equation in terms of weight per gram of dry matter. W5h is the leaf weight of leaves after 5 h in darkness and 20 °C, W24h is the weight of the leaves isolated after 24 h in darkness and 20 °C and DW is the dry leaf weight (48 °C at 70 °C). The cuticle transpiration was calculated using the equation [25]:

$$CT = (W5h - W24h)/DW \qquad (5)$$

Canopy temperature depression (CTD) was determined by measurements with a hand-held infrared thermometer (Raytek Raynger ST20 Infrared Thermometer, Santa Cruz, CA, USA). A few days after irrigation, canopy temperatures (CT) were measured between 12:00 and 1:30 pm on cloudy and sunny days. For this experiment four measurement points for plant canopy temperature were chosen in each pot at approximately 15–30 cm above the leaves of the strawberry plants, approximately 30–60° from the horizontal position. Ambient temperatures (AT) were measured with a thermometer held at plant height. CTD was worked calculated following [26]:

$$CTD = AT - CT \qquad (6)$$

Epicuticular wax layer (EWL): for determining EWL, the method of Ebercon et al. [19] was used. This measure is based on the color change that occurs when acidic potassium dichromate ($K_2Cr_2O_7$) reacts with epicuticular wax. Two fully expanded leaves were harvested from each plant in each pot (six leaf disks in each replication including three biological replicates). Leaf disks (5.699 cm^2) were isolated by hole-punch, and used for wax extraction. These disks were put in a tube and 15 mL of chloroform was added and the tube shaken at room temperature for 15 s. The extract was evaporated to dry in a water bath maintained at 90 °C. Then, five ml of the $K_2Cr_2O_7$ solution was added to the tube and the reaction mixture left in a boiling water bath for 30 min. When the samples were cooled 10 mL of distilled water was added to tubes, tubes were mixed and finally, the absorbance was measured at 590 nm using a Spectrophotometer (Ltd T80 + UV/VIS; PG Instruments, Leicestershire, UK). The standard curve calibration was produced by using known concentrations of polyethylene glycol-6000 for EWL determination at 590 nm wavelength [19].

Scanning electron microscope (SEM) images were captured and used to examine differences in wax morphology. Preparation of leaf samples followed the method reported by Åström et al. [27]. The youngest fully developed leaf after the end of fruit production was harvested. The leaf pieces were cut from the central part of the middle leaflet, near the widest point of each leaf. The samples were fixed individually in FAA (formalin acetic acid-alcohol) solution (36% paraformaldehyde, 100% acetic acid, 85% ethanol; 10:5:85 by volume) for a minimum of 3 weeks. After fixation, the samples were dehydrated through an ethanol series (25%, 50%, 75%, and 100%) [27]. 5-8 mm completely dried pieces of prepared leaf samples, were attached with double adhesive tape to the aluminum stubs and sputter-coated with gold particles. Coated surfaces were observed using a Philips Xl 30 scanning electron microscope (Philips XL30 SEM, Amsterdam, The Netherlands) at an accelerating voltage of 10 kV [28]. SEM images of epicuticular wax of strawberry leaves at two levels of magnification (Bars; 100 μm and 25 μm) were taken at the University of Guilan.

The leaf free proline content for the strawberry plants was extracted and determined by following the method described by Bates et al. [29]. 500 mg of the leaf samples were homogenized in 5 mL sulfosalicylic acid (3%) and the homogenate centrifuged at 3500× *g* for 10 min. The supernatant was mixed with 2 mL acid ninhydrin [1.25 g of ninhydrin in glacial acetic acid (30 mL) and 6 M phosphoric acid (20 mL), with agitation, which was warmed until dissolved for Acid ninhydrin preparation] and 2 mL of glacial acetic acid in a test tube at 100 °C for 60 min, and the reaction terminated in an ice bath. The reaction mixture was extracted with 4 mL toluene, mixed vigorously with a test tube stirrer for 15–20 s. Free proline content was quantified spectrophotometrically at 520 nm using L-proline as a standard. The absorbance was measured at 520 nm. The content of proline was determined using a standard curve and expressed as μmol g^{-1} fresh weight following [29].

Pigment parameters of the leaves including chlorophyll and carotenoid content were measured following a method described by Abdi et al. (2016). Initially 500 mg of leaf tissues were placed in each tube with 50 mL 80% acetone solution, these samples were homogenized and then the extract sap was centrifuged for 10 min at 3000× *g* and absorbance of the supernatant measured at 663 nm (for chlorophyll a), 645 nm (for chlorophyll b), and 470 nm (for total carotenoids). Finally, the pigment content was calculated according to the following formulas [21]:

$$\text{Chl a} = 11.75 \times A662 - 2.35 \times A645 \tag{7}$$

$$\text{Chl b} = 18.61 \times A645 - 3.96 \times A662 \tag{8}$$

$$\text{Car} = 1.000 \times A470 - 2.27 \times \text{Chl a} - 81.4 \times \text{Chl b}/227 \tag{9}$$

2.3. Statistical Analysis

The plants were arranged in a Completely Randomized Design in a factorial layout with three factors: Salt (0, 25 and 50 mM), nano-silicon dioxide concentrations (0, 50 and 100 mg L^{-1}) before flowering and two levels of nano-silicon dioxide (0 and 50 mg L^{-1}) after flowering, with three replications and 12 pots (plants) per replication. All data were analyzed by a one-way analysis of variance and mean comparisons were made by least significant differences (LSD) with software (SAS, v. 9.4, Cary, NC, USA).

3. Results

Salinity and nano-silicon dioxide treatments resulted in changes in strawberry plant growth characteristics; for example the 100 mM salt treatments resulted in decreases in root and shoot fresh weight (by 35 and 65%, respectively) and in root and shoot dry weight (by 50% in the shoot to root ratio and 26% in root volume; Table 2). As expected, the 50 mM NaCl treatments reduced these growth characteristics more than the 25 mM NaCl treatment (Table 1).

Table 2. Effect of $nSiO_2$ and salt stress on biomass parameters, root and shoot dry weight and fresh weight of strawberry cv "Camarosa", including analysis of variance.

	Root Fresh Weight (g)	Root Dry Weight (g)	Shoot Fresh Weight (g)	Shoot Dry Weight (g)	Shoot/Root	Root Volume (cm^3 per plant)
Salinity (mM)						
0	52.1 a	10.99 a	51.94 a	16.16 a	0.996 a	51.00 a
25	43.2 b	8.44 b	36.76 b	12.23 b	0.850 a	41.88 b
50	36.7 c	7.11 c	18.14 c	5.58 c	0.494 b	37.50 b
Nano-silicon Dioxide (mg L^{-1})						
S_1	36.72 c	6.95 c	28.54 c	9.22 c	0.777 a	35.88 b
S_2	44.86 abc	9.04 ab	33.97 b	10.7 bc	0.757 a	41.77 ab
S_3	50.92 a	10.35 a	38.33 b	11.50 ab	0.752 a	49.77 a
S_4	41.41 bc	8.01 bc	35.92 b	11.25 abc	0.867 a	41.11 ab
S_5	46.34 abc	9.13 ab	35.74 c	12.24 ab	0.771 a	43.77 ab
S_6	47.53 ab	9.92 a	41.18 a	13.04 a	0.866 a	48.44 a
Analysis of Variance						
Salinity	**	**	**	**	**	**
$nSiO_2$	**	*	*	**	ns	ns
Salinity × $nSiO_2$	ns	ns	ns	ns	ns	ns

Means of the main effects followed by different letters in each column indicate significant difference at $p \leq 0.05$ by the least significant difference (LSD). ns, * or ** indicate non-significance ($p > 0.05$) or significance at $p \leq 0.05$ or $p \leq 0.01$, by the F-test, respectively.

Incorporation of nano-silicon dioxide ($nSiO_2$) into the nutrient solution changed some of the growth parameters measured for the strawberry plants. For example, the plants treated with $nSiO_2$ had higher root fresh and dry weight as compared to 0 mg L^{-1} $nSiO_2$ under salt stress conditions (Table 2). The highest root dry weight (10.4 g) and fresh weight (50.9 g) was observed when plants were treated with 50 mgL^{-1} $nSiO_2$ before full flowering (Si_3).

Shoot fresh and dry weight were significantly affected individually by salinity and $nSiO_2$ treatments, but no significant difference was found for any interaction effect of salinity and $nSiO_2$ (Table 2). Strawberry plants which received 100 mg L^{-1} $nSiO_2$ before the flowering stage and 50 mg L^{-1} thereafter (Si_6) showed the highest fresh shoot weight (41.2 g), while the highest shoot dry weight (13 g) was recorded for plants which received 100 mg L^{-1} $nSiO_2$ before flowering and 50 mg L^{-1} thereafter (Si_6) or plants that received 50 mgL^{-1} $nSiO_2$ before flowering stage (Si_3) (Table 2); and differences between $nSiO_2$ treated and control (no $nSiO_2$) plants for shoot to root ratio and root volume were also recorded (Table 2). A t-test was conducted to explore any differences between the addition (S_6) and absence (S_1) of silicon in the nutrient solution under salinity stress conditions. This revealed that there were differences in the epicuticular wax layer and proline (Table 3).

Table 3. Student's *t*-test of nano-silicon dioxide effects on morphological and physiological parameters of strawberry plants exposed to 50 mM NaCl salinity stress; ns (no significant difference); Pr > [t] (*p*-value for the effect of the variable on the response and *t* statistic) * (significant difference).

	S_1			S_6			T Value	Pr > F	Pr > [t]	
	Mean	Std Dev	Std Err	Mean	Std Dev	Std Err			Pooled (Equal)	Satterthwaite (Unequal)
Fresh weight	12.98	0.849	0.49	21.32	3.89	2.24	−3.63	0.0909	0.0222 *	0.059 ns
Dry weight	4.28	1.017	0587	6.55	0.606	0.35	−3.31	0.524	0.0297 *	0.0402 *
Root fresh weight	30.42	4.79	2.76	34.55	7.72	4.45	−0.79	0.475	0.475 ns	0.483 ns
Root dry weight	4.92	0.70	0.407	8.29	2.314	1.33	−2.41	0.17	0.0733 ns	0.117 ns
Root volume	28.33	7.63	4.40	40.00	5.00	2.88	−2.21	0.60	0.091 ns	0.102 ns
Shoot/root	0.431	0.0449	0.0259	0.637	0.183	0.106	−1.89	0.112	0.131 ns	0.185 ns
Membrane stability index (MSI)	64.22	10.31	5.95	80.00	2.68	1.55	−2.57	0.127	0.062ns	0.109 ns
Proline	13.42	0.549	0.316	8.19	0.641	0.370	10.72	0.844	0.0004 **	0.0005 **
Epicuticular wax layer (EWL)	17.06	5.65	3.266	34.03	8.29	4.78	−2.93	0.635	0.043 *	0.050 *

A significant difference was found for the individual effects of salinity and nSiO$_2$ treatments on strawberry fruit yield but there was no significant difference for any interaction effects on fruit yield (Table 4). The lowest fruit yield was observed when plants were treated with 50 mM NaCl as compared to controls (no NaCl), as the salt treatment decreased fruit yield by 61%. Furthermore, application of nSiO$_2$ led to an overall improvement in fruit yield. The highest fruit production per plants (161 g) was obtained when plants received 100 mgL^{-1} nSiO$_2$ before flowering and 50 mg L^{-1} after flowering stage (Si$_6$) (Table 4).

Table 4. Effect of nSiO$_2$ and salt stress on fruit yield, Relative Water Content (RWC); Relative Water Protection (RWP); Relative Water Loss (RWL); Membrane Stability Index (MSI), Cuticle Transpiration (CT) and canopy temperature for strawberry cv 'Camarosa'.

	Fruit Yield (g)	RWC (%)	RWP (%)	RWL (%)	MSI (%)	CT (g H$_2$O/g Dry Weight)	Canopy Temperature (°C)	
							Cloudy Day	Sunny Day
Salinity (mM)								
0	198.06 a	85.1 a	0.91 a	0.156 a	83.9 a	0.587 a	3.91 a	2.72 a
25	149.40 b	81.79 a	0.87 ab	0.154 a	79.3 a	0.832 a	3.57 a	2.04 a
50	77.39 c	67.37 b	0.86 b	0.168 a	75.5 b	0.908 a	2.18 b	0.10 b
Nano-silicon dioxide (mg L^{-1})								
S_1	124.05 c	76.71 a	0.861 a	0.107 c	74.2 bc	1.111 a	2.33 c	0.713 c
S_2	142.33 abc	75.99 a	0.901 a	0.161 ab	79.1 abc	0.593 a	3.15 b	0.861 c
S_3	151.92 ab	75.26 a	0.877 a	0.161 ab	79.6 abc	0.788 a	3.20 b	1.14 c
S_4	140.79 bc	78.33 a	0.883 a	0.202 a	84.64 a	0.753 a	3.15 b	1.46 bc
S_5	129.34 c	80.91 a	0.875 a	0.158 b	71.9 c	0.843 a	3.48 ab	3.06 a
S_6	161.26 a	81.32 a	0.911 a	0.168 ab	82.2 ab	0.564 a	4.02 a	2.48 ab
Analysis of Variance								
Salinity	**	**	ns	ns	**	ns	**	**
Nano-silicon dioxide	**	ns	ns	**	*	ns	**	**
Salinity × Naon-silicon dioxide	ns	ns	ns	ns	ns	ns	ns	ns

Means of the main effects followed by different letters in each column indicate significant difference at $p \le 0.05$ least significant range (LSD). ns, * or ** indicate non-significance ($p > 0.05$) or significance at $p \le 0.05$ or $p \le 0.01$, by the F-test, respectively.

Physiological parameters such as RWC, RWP, and MSI significantly decreased, when strawberry plants were exposed to salinity [reduced by 21%, 5.5% and 10% relative to measures in control (no NaCl) plants, respectively], but RWL was not affected by salt stress. The lowest values were recorded for plants were exposed to 50 mM NaCl (Table 4).

There was no significant difference between $nSiO_2$ treatments and control (no $nSiO_2$) for RWC and RWP, but RWL and MSI of $nSiO_2$ treated plants was significantly higher than control (no $nSiO_2$) plants. The highest RWL and MSI was measured in plants that continuously received 50 mg L^{-1} $nSiO_2$ (Si_4) over the growth and development stages (Table 4).

The canopy temperature of strawberry plants was significantly reduced by salt stress, especially when the plants had been exposed to 50 mM NaCl during growth and development. Nano-silicon dioxide application raised canopy temperature of strawberry plants both in cloudy and sunny days (Table 4). No significant difference was observed for cuticle transpiration (CT) in $nSiO_2$ treated and control (no $nSiO_2$) plants.

Proline content of salt treated strawberry plant leaves increased by 15 and 81% under 50 mM and 100 mM salinity treatments but incorporation of $nSiO_2$ to the nutrient solution limited proline accumulation. The highest proline content was found in 0 mg L^{-1} $nSiO_2$ (S_1) treated plants under salt stress conditions (Table 5; Figure 1). NSiO2 treatment caused a significant decrease in proline content in salt stress plants compared to the strawberry plants treated with salt treatments without nano-silicon dioxide treatment. The results revealed a negative correlation (-0.63058 **; $p < 0.01$) between proline content and EWL.

There were differences in the epicuticular wax layer and proline content of salt and $nSiO_2$ treated plants (Figures 1 and 2). The epicuticular wax layer (EWL) was significantly reduced in strawberry plants when exposed to salt stress relative to controls (no NaCl). EWL was low when plants were exposed to 25 and 50 mM NaCl compared to controls (no NaCl) (Table 4). $NSiO_2$ treated plants had higher EWL than controls (no $nSiO_2$). The highest EWL observed was in plants that received 100 mgL^{-1} $nSiO_2$ before flowering and 50 mg L^{-1} thereafter (Si_6).

Table 5. Effect of nSiO2 and salt stress on Epicuticular Wax Layer (EWL), proline, chlorophyll (Chl a and Chl b and total) and carotenoids content of strawberry cv Camarosa under various conditions tested.

	EWL (μg cm^2)	Proline (μmol g^{-1})	Chl a (mg g^{-1} Fresh Weight)	Chl b (mg g^{-1} Fresh Weight)	Total Chl (mg g^{-1} Fresh Weight)	Carotenoids (mg g^{-1} Fresh Weight)
			Salinity (mM)			
0	63.43 a	5.83 c	7.78 a	2.75 a	10.53 a	2.86 b
25	36.52 b	6.68 b	7.41 b	2.88 a	10.30 a	3.24 a
50	28.54 b	10.53 a	5.96 c	2.38 b	8.35 b	2.63 c
			Nano-Silicon Dioxide (mg L^{-1})			
S_1	35.53 b	8.36 a	6.48 c	2.38 c	8.86 c	2.66 c
S_2	43.74 ab	7.06 bcd	6.87 bc	2.46 c	9.34 bc	2.69 c
S_3	45.89 ab	7.69 abc	6.68 c	2.39 c	9.08 c	2.82 bc
S_4	43.22 ab	7.91 ab	7.57 a	3.11 a	10.68 a	3.02 ab
S_5	42.57 ab	6.03 d	7.53 a	2.96 a	10.50 a	3.23 a
S_6	47.12 a	6.61 cd	7.18 ab	2.71 b	9.89 b	3.03 ab
			Analysis of Variance			
Salinity	**	**	**	**	**	**
Nano-silicon dioxide	ns	**	**	**	**	**
Salininty × Nano-silicon dioxide	*	*	**	**	**	**

Means of the main effects followed by different letters in each column indicate significant difference at $p \leq 0.05$ by least significant range (LSD). ns, * or ** indicate non-significance ($p > 0.05$) or significance at $p \leq 0.05$ or $p \leq 0.01$, by the F-test, respectively.

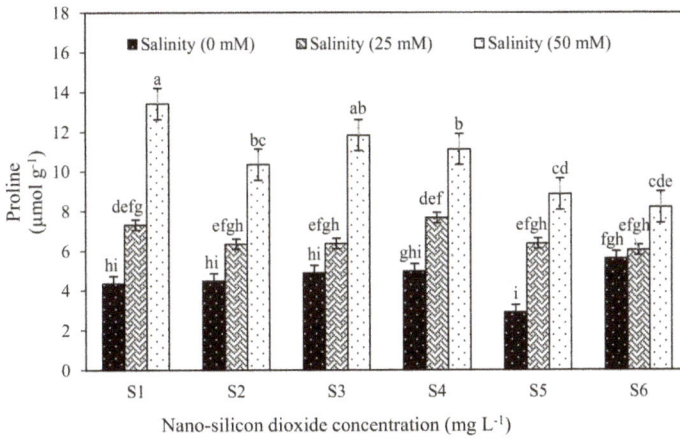

Figure 1. Proline concentrations of strawberry leaves from plants grown in three levels of salinity 0 mM (black bars), 25 mM (grey bars) and 50 mM (white bars) and treated with different levels of nano-silicon dioxide. Mean values with the same letters are not significantly different by least significant differences (LSD) test at $p \leq 0.01$. The content of photosynthetic pigments such as chlorophylls and carotenoids was significantly reduced in salt stressed plants relative to controls (no NaCl), especially for the 50 mM NaCl treatment where there was a 21% decrease in the total chlorophyll. Photosynthetic pigment content, including chlorophyll a, decreased in response to the salinity stress treatments and in contrast, the chlorophyll b and carotenoid content increased in response to the mild salinity level, but under the more severe 50 mM NaCl stress these pigments were reduced in comparison to controls (no NaCl). The treatments with $nSiO_2$ increased chlorophyll a, b and total chlorophyll and carotenoid content compared to controls (no $nSiO_2$) under stress and non-stress condition (Table 5).

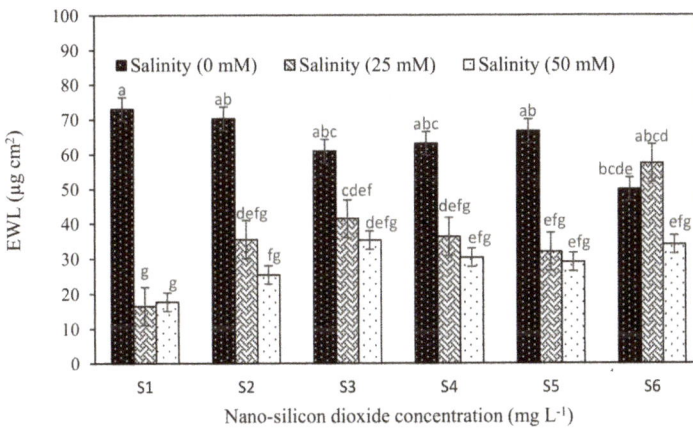

Figure 2. EWL concentrations of strawberry in three levels of salinity 0 mM (black bars), 25 mM (grey bars) and 50 mM (white bars) treated with different levels nano-silicon dioxide. Mean values with the same letter are not significantly different by least significant differences (LSD) test at $p \leq 0.01$.

To further investigate the quantitative differences in EWL (Table 4; Figure 2) imaging techniques were used to check for qualitative differences EWL (Figures 3 and 4). The interaction effect of $nSiO_2$ and salinity on EWL had revealed that salinity treatments (25 and 50 mM) significantly reduced EWL both in control (no $nSiO_2$) and $nSiO_2$ treated strawberry plants except for the plants pre-treated with 100 mg L^{-1} $nSiO_2$ before BBCH: 61 and 50 mg L^{-1} after BBCH: 61. This treatment increased EWL under salinity stressed conditions especially when plants were exposed to moderate stress (Figures 2–4). The scanning electron microscopic (SEM) images revealed that there were two forms of wax crystals on the strawberry leaf surface; regular (rougher) like a spider web structure and irregular (smoother) crystals. In the non-stressed conditions, leaf surfaces were covered with irregular-shaped wax crystals and formed a dense network. The size of the wax crystal was thicker and a less dense network was observed in plants treated with NaCl in comparison to controls (no NaCl). The crystal was deposited in the epicuticle layer when plants were treated with $nSiO_2$, and this appeared to result in an increase in thickness in the wax crystal under stress conditions. Additionally, a crystal structure with sparser arrangements of plate-shaped wax under salt stress conditions was observed, suggesting a decrease in the total number of crystalloids present per unit area compared to control (no NaCl). Notable changes in wax morphology occurred in plants treated with 50 mM NaCl. Overall, the results clearly showed that as salinity increased epicuticular wax crystals, displayed morphology changes at the strawberry leaf surface (Figures 3 and 4).

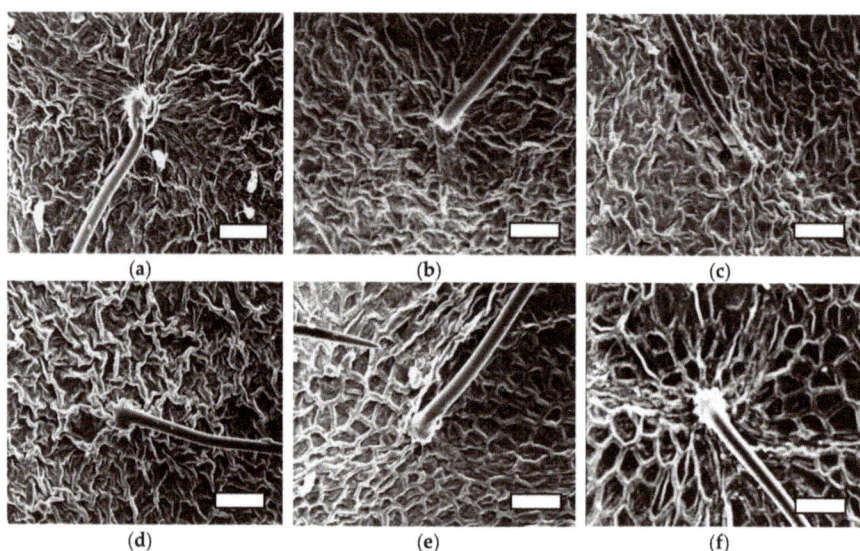

Figure 3. The effects of salt stress and $nSiO_2$ on the epidermal cell walls of strawberry leaves. (**a**) Cont; 0 mM NaCl + 0 mg L^{-1} $nSio_2$, (**b**) 0 mM NaCl + 100.50 mg L^{-1} SiO_2 (**c**) 25 mM NaCl + 0 mg L^{-1} SiO_2 (**d**) 25 mM NaCl + 100,50 mg L^{-1} SiO_2, (**e**) 50 mM NaCl + 0 mg L^{-1} SiO_2, (**f**) 50 mM NaCl + 100.50 mg L^{-1} SiO_2. Scanning electron microscope (SEM) image of the strawberry leaves. White scale bars = 100 μm.

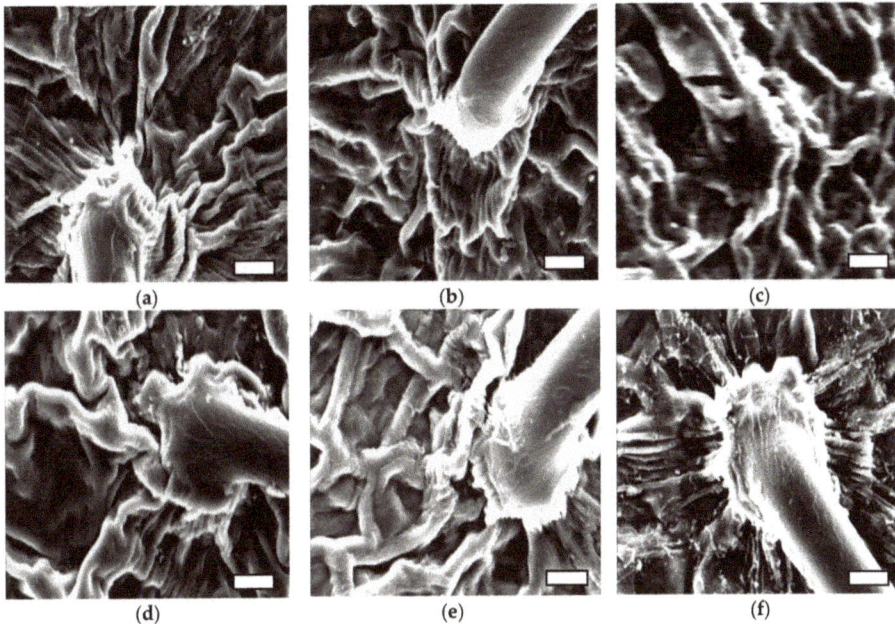

Figure 4. The effects of salt stress and nSiO2 on the epidermal cell walls of strawberry leaves. (a) Cont; 0 mM NaCl + 0 mg L−1 nSio2, (b) 0 mM NaCl + 100,50 mg L−1 SiO2 (c) 25 mM NaCl + 0 mg L−1 SiO2 (d) 25 mM NaCl + 100,50 mg L−1 SiO2, (e) 50 mM NaCl + 0 mg L−1 SiO2, (f) 50 mM NaCl + 100,50 mg L−1 SiO2. Scanning electronmicroscope (SEM) image of the strawberry leaves. White scale bars = 20 μm.

4. Discussion

Differences in morphological, physiological and biochemical characteristics, such as shoot and root fresh weight and dry weight, RWC, EWL, RWL, cuticle transpiration, and MSI, and proline content and canopy temperature were observed in strawberry plants treated with different combinations of salinity and nano-silicon dioxide treatments. Application of $nSiO_2$ reduced the negative effects of salinity and improved vegetative growth of strawberry plants (Tables 2–5). These findings are consistent with previous reports for similar studies in other plant species (Figure 5), which demonstrated that $nSiO_2$ increased proliferation of apple (*Malus pumila* Borkh) explants under non-stressed or osmotic-stressed conditions [21,30]. The application of silicon was also shown to increase root growth of rice (*Oryza sativa* L.) plants under drought stress conditions [31]; and an increase in Si-mediated root growth was observed in sorghum (*Sorghum bicolor*) under drought stress [32]. However, root growth recovery with silicon treatments after salt stress has not always been observed. For example, positive effects have been reported for silicon treatments, on the shoot growth of wheat (*Triticum aestivum* L.), but without obvious effect on the roots [33]; and similar observations were made for cucumber plants [34].

The beneficial effect of $nSiO_2$ in relation to improving germination of soybean (*Glycine max*) seeds was suggested to be related to increasing nitrate reductase activity [35], and was linked to plants ability to uptake and use water and nutrition by seeds [36]. Another suggestion as to the benefits of silicon or nano-silicon for plants grown under stressful conditions relates to increased photosynthetic rate, stomatal conductance, and water use efficiency; traits which then improve the tolerance to salinity of tomato plants [37]. Here we observe and explore a possible association between $nSiO_2$ treatment, epicuticular wax, and proline accumulation.

Proline is an osmolyte that usually accumulates under stress conditions and plays an important role in osmotic adjustment in plants [38]. It has been reported that the proline content of wheat leaves increased under water stress conditions, while the addition of silicon decreased proline accumulation, consistent with proline accumulation being linked as a sign of stress damage in experimental conditions [39]. The proline content in sorghum plants under drought stress conditions decreased significantly, while sugar contents in the roots were reported to be increased by silicon treatments [32]. Si application in soybean plants has been reported to cause a reduction in proline content under drought stress [40]. Proline has been considered as a possible carbon and nitrogen source for rapid recovery from stress, a stabilizer for membranes and some macromolecules, and also a free radical scavenger [41]. For example proline content increased in maize seedlings when exposed to salinity treatments, but decreased with Si plus NaCl treatments [42]; and in this example Si may provide a protective role helping to prevent lipid peroxidation induced by NaCl, because this was significantly lower in the Si-treated maize seedlings under salt stress than those under salt stress without Si treatment [42]. Both epicuticular wax and proline content have been reported to be significantly increased during water deficit conditions [43]. But in the current study, proline content increased and epicuticular wax (EWL) decreased in strawberry plants when exposed to salt stress. In this study moderate salt stress (25 mM) when followed by 100 mg L^{-1} nSiO$_2$ before BBCH: 61 and 50 mg L^{-1} after BBCH: 61 significantly increased EWL (Table 5; Figure 2).

The role of silicon in regulating the water status of plants is of interest, particularly in the context that the initial reduction of the growth of plants under salt stress is due to the osmotic effect of the salt [44]. The researchers found that RWC increased in response to silicon treatments under stress conditions, not only by reducing transpiration rate through the deposition of silicon in leaf and stem epidermis cells, but also by increasing potassium absorbance and translocation to stomatal guard cells, where potassium influences stomatal conductivity [45,46]. It has been suggested that Si can increase plants water content under salinity stress, due to findings that Si reduced the osmotic potential (more negative) and increased turgor pressure of tomato leaves under salt stress [47]. In this study, RWC of strawberry plants when treated with nSiO$_2$ was higher than that of control plants (Table 3). This observation is consistent with previous reports indicating that RWC in wheat plants was significantly lower under stress conditions, and adding silicon nutrition completely restored RWC to the levels observed in the non-stress plants [48,49]. Similar effects of silicon on RWC of leaf beans were reported for plants grown in hydroponic culture [50]. Overcoming the osmotic stress and physiological deficiency of conditions where water is limited is one of the most important adaptation strategies of plants under salinity conditions. The research on the influence of silicon has shown that it can significantly improve the outcome for salt stressed plants, and the mode of action may be in preventing the loss of water from plants by reducing the rate of transpiration [51].

In the current work, we observed that the membrane stability index (MSI) was markedly decreased by salt stress. Previously it has been shown that strawberry plants under salinity accumulated more H$_2$O$_2$ compared to control plants, and a combination of salinity with silicon nutrition via the nutrient solution significantly ameliorated the impact of salinity on membrane integrity, lipid peroxidation and H$_2$O$_2$ content [52].

The results of this study showed that canopy temperature of strawberry plants increased under salt stress. Previous studies revealed that lower canopy temperature genotypes appear to exhibit better tolerance to drought stress; for example, in water stress conditions, increases in canopy temperature were observed in wheat (*Triticum aestivum* L. and *T. durum* L.) [53] and cowpea (*Vigna unguiculata* L.) varieties [54]. We suggest that for strawberry plants limited water availability under salt stress conditions results in rising canopy temperatures. Given the fact that the temperature, amount of light and moisture content affect the morphology of leaf wax, and since variability in these factors coincide, it is difficult to detect their individual effects [55]. However, plants with well-developed layers of epicuticular wax showed lower leaf and canopy temperatures, reduced rates of transpiration, and improved water status as compared to control, and also plants adapted to hot climatic conditions possess a thick cuticle with reduced transpiration rates [56].

Silicon application may reduce the loss of water through the cuticle due to silica deposition underlying epidermal cells of leaf and stem plants influencing water loss [46]. The formation of a silica-cuticle double layer on leaf epidermal cells may be effective in altering leaf transpiration and water loss from the leaf surface could be limited due to Si deposition [57]. Si accumulates in the epidermal tissues, and a layer of cellulose matrix-Si is created when calcium and pectin are present, which provides protection to the plant [58]. Silicification occurs in the endodermis in parts of roots of gramineae during maturation; and in the cell walls of other tissues including vascular, epidermal, and cortical cells in older roots; and in shoots including hull and leaf sheath, as well as in the inflorescence [58].

Results of an investigation on rice showed that a layer of deposited Si (2.5 μm) is formed under cuticle with a double layer of Si-cuticle in the leaf blades [59]. Results of other studies revealed that the silicified structures were found on cell wall epidermal surfaces as separate rosettes and knobs sheltered in spicules, also silicon deposition on surfaces has effects on stem (3–7 mm) and leaf (0.2–1.0) thickness [58]. Silicon is absorbed in roots, and transported passively through the transpiration stream and deposited in beneath the cuticle, forming a cuticle-silica double layer [60]. It was suggested that this physical barrier delayed and reduced the penetration of fungus in rice leaves, cucumber, melon (*Cucumis melo* L) and pumpkin (*Cucurbita pepo* L.), and vine seedlings [61].

In the current study, nSiO$_2$ application was associated with strawberry plants maintaining higher chlorophyll content under saline conditions. Therefore, addition of nSiO$_2$ in nutrient solutions could help alleviate the negative effects of NaCl on chlorophyll content in strawberry. Inhibition of chlorophyll biosynthesis, and acceleration of its degradation and oxidative damage induced by salinity could be considered as reasons for the declining chlorophyll content [52]. Further research is needed to explore the influence of silicon on the biosynthesis of new chlorophyll and the protection of existing chlorophyll against salinity-induced oxidative stress [52]. Previous studies showed that salt stress in cowpea, kidney bean (*Phaseolus vulgaris* L.), faba bean (*Vicia faba* L.) and soybean caused significant reductions in plant growth, but Si supplementation greatly improved the growth of these plants by increasing total photosynthetic pigments and photosynthetic rate, chlorophyll content, stomatal conductance, transpiration, and intercellular carbon dioxide concentration [62].

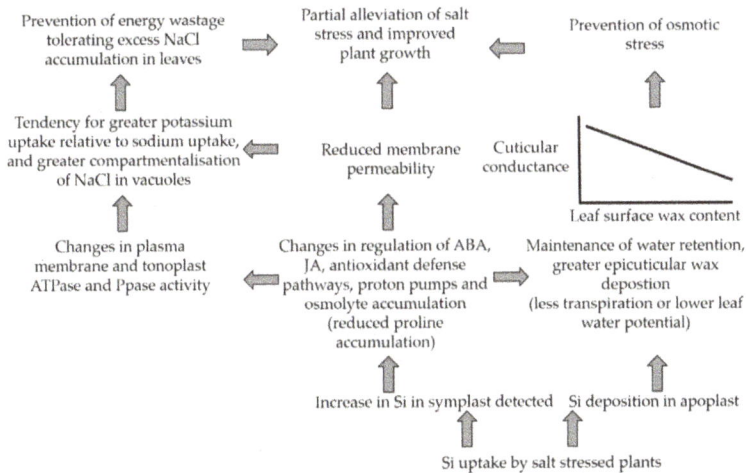

Figure 5. Schematic diagram indicating the beneficial responses that occur in salt stressed plants when they are supplied supplemental silicon. Abscisic acid (ABA); jasmonic acid (JA), ATPase (enzymes that catalyze the decomposition of ATP into ADP and a free phosphate ion or the inverse reaction), Ppase (proton-pumping pyrophosphatase); schematic adapted from Liang et al. 2015 [46] and the representation of the relationship between cuticular conductance and leaf wax surface content is adapted from Agarie et al., 1998 [63].

5. Conclusions

Salinity stress treatments were detrimental to morphological and physiological parameters of strawberry plants. In this study, nSiO$_2$ treatments suppressed the negative effects of salinity, possibly by improving the Epicuticular Wax Layer (EWL); and nSiO$_2$ treatments enabled salt stressed plants to better maintain their chlorophyll content and leaf relative water content (RWC) and relative water protection (RWP) relative to controls (no SiO$_2$). Observations were made that are relevant to improving strawberry productivity in both saline and control (no added NaCl) conditions, in particular, the data indicated that application of 50 mg L^{-1} nSiO$_2$ before stage 'BBCH:61' increased root growth, and that treatments with 100 mg L^{-1} nSiO$_2$ positively influenced strawberry plant growth rate and productivity (Table 2). We conclude by suggesting three possible directions for future research: (1) Further exploring how variation in the timing of silicon treatments influences EWL deposition by testing EWL at multiple plant developmental stages; (2) investigation of whether there is genetic variation for EWL deposition in strawberry; and (3) testing to distinguish the benefit of greater EWL deposition in saline conditions relative to the benefit of the other signalling and physiological changes that are linked to increased silicon uptake (Figure 5).

Author Contributions: Conceptualization, S.A., M.G., and M.E.; Methodology, S.A.; Formal analysis, S.A., M.G., M.E., C.S.B.; Investigation, S.A.; Data curation, S.A.; Writing—Original draft preparation, S.A.; Writing—Review and editing, S.A. and C.S.B.; Supervision, M.G. and M.E.

Funding: This research was funded by Iran Nanotechnology Innovation Council (INIC) under the grant number of 116399, the University of Guilan under the grant number of 1397/2690481881 and also partially was funded by Hasan Ebrahimzade Maboud's Charity Fund.

Conflicts of Interest: The authors declare no conflict of interest. The funders had no role in the design of the study; in the collection, analyses, or interpretation of data; in the writing of the manuscript, or in the decision to publish the results.

References

1. Gómez-del-Campo, M.; Baeza, P.; Ruiz, C.; Lissarrague, J.R. Water-stress induced physiological changes in leaves of four container-grown grapevine cultivars (*Vitis vinifera* L.). *VITIS J. Grapevine Res.* **2015**, *43*, 99.
2. Azizinya, S.; Ghanadha, M.R.; Zali, A.A.; Samadi, B.Y.; Ahmadi, A. An evaluation of quantitative traits related to drought resistance in synthetic wheat genotypes in stress and non-stress conditions. *Iran. J. Agric. Sci.* **2005**, *36*, 281–293.
3. Malakuti, M.J.; Keshavarz, P.; Karimian, N. *Comprehensive Diagnosis and Optimal Fertilizer Recommendation for Sustainable Agriculture*; Tarbiat Modares University Press: Tehran, Iran, 2008; p. 132.
4. Sun, Y.; Niu, G.; Wallace, R.; Masabni, J.; Gu, M. Relative Salt Tolerance of Seven Strawberry Cultivars. *Horticulturae* **2015**, *1*, 27–43.
5. Rousseau-Gueutin, M.; Lerceteau-Köhler, E.; Barrot, L.; Sargent, D.J.; Monfort, A.; Simpson, D.; Arús, P.; Guérin, G.; Denoyes-Rothan, B. Comparative genetic mapping between octoploid and diploid fragaria species reveals a high level of colinearity between their genomes and the essentially disomic behavior of the cultivated octoploid strawberry. *Genetics* **2008**, *179*, 2045–2060. [CrossRef] [PubMed]
6. Flam-Shepherd, R.; Huynh, W.Q.; Coskun, D.; Hamam, A.M.; Britto, D.T.; Kronzucker, H.J. Membrane fluxes, bypass flows, and sodium stress in rice: The influence of silicon. *J. Exp. Bot.* **2018**, *69*, 1679–1692. [CrossRef]
7. Bao-Shan, L.; Shao-Qi, D.; Chun-Hui, L.; Li-Jun, F.; Shu-Chun, Q.; Min, Y. Effect of TMS (nanostructured silicon dioxide) on growth of Changbai larch seedlings. *J. For.* **2004**, *15*, 138–140. [CrossRef]
8. Alva, A.K.; Mattos, D.; Paramasivam, S.; Patil, B.; Dou, H.; Sajwan, K.S. Potassium Management for Optimizing Citrus Production and Quality. *Int. J. Fruit Sci.* **2006**, *6*, 3–43. [CrossRef]
9. Gao, X.; Zou, C.; Wang, L.; Zhang, F. Silicon Decreases Transpiration Rate and Conductance from Stomata of Maize Plants. *J. Plant Nutr.* **2006**, *29*, 1637–1647. [CrossRef]
10. Gao, X.; Zou, C.; Wang, L.; Zhang, F. Silicon Improves Water Use Efficiency in Maize Plants. *J. Plant Nutr.* **2005**, *27*, 1457–1470. [CrossRef]
11. Naranjo, E.M.; Andrades-Moreno, L.; Davy, A.J. Silicon alleviates deleterious effects of high salinity on the halophytic grass Spartina densiflora. *Plant Physiol. Biochem.* **2013**, *63*, 115–121. [CrossRef] [PubMed]

12. Ahmed, M.; Kamran, A.; Asif, M.; Qadeer, U.; Ahmed, Z.I.; Goyal, A. Silicon priming: A potential source to impart abiotic stress tolerance in wheat: A review. *Aust. J. Crop Sci.* **2013**, *7*, 484.

13. Shi, Y.; Wang, Y.; Flowers, T.J.; Gong, H. Silicon decreases chloride transport in rice (Oryza sativa L.) in saline conditions. *J. Plant Physiol.* **2013**, *170*, 847–853. [CrossRef]

14. Agarie, S. Effect of silicon on growth, dry matter production and photosynthesis in rice plants. *Crop Prod. Improv. Tech. Asia* **1993**, 225–234.

15. Prasad, T.; Sudhakar, P.; Sreenivasulu, Y.; Latha, P.; Munaswamy, V.; Reddy, K.R.; Sreeprasad, T.S.; Sajanlal, P.R.; Pradeep, T. Effect of nanoscale zinc oxide particles on the germination, growth and yield of peanut. *J. Plant Nutr.* **2012**, *35*, 905–927. [CrossRef]

16. Wang, L.-J.; Wang, Y.-H.; Li, M.; Fan, M.-S.; Zhang, F.-S.; Wu, X.-M.; Yang, W.-S.; Li, T.-J. Synthesis of ordered biosilica materials. *Chin. J. Chem.* **2002**, *20*, 107–110. [CrossRef]

17. Samuels, A.L. The Effects of Silicon Supplementation on Cucumber Fruit: Changes in Surface Characteristics. *Ann. Bot.* **1993**, *72*, 433–440. [CrossRef]

18. Jenks, M.A.; Ashworth, E.N. Plant epicuticular waxes: Function, production, and genetics. *Hortic. Rev.* **1999**, *23*, 1–68.

19. González, A.; Ayerbe, L. Effect of terminal water stress on leaf epicuticular wax load, residual transpiration and grain yield in barley. *Euphytica* **2010**, *172*, 341–349. [CrossRef]

20. Cameron, K.D.; Teece, M.A.; Smart, L.B. Increased accumulation of cuticular wax and expression of lipid transfer protein in response to periodic drying events in leaves of tree tobacco. *Plant Physiol.* **2006**, *140*, 176–183. [CrossRef]

21. Abdi, S.; Abbaspur, N.; Avestan, S.; Barker, A.V. Physiological responses of two grapevine (*Vitis vinifera* L.) cultivars to Cycocel™ treatment during drought. *J. Hortic. Sci. Biotechnol.* **2016**, *91*, 211–219. [CrossRef]

22. Hasheminasab, H.; Assad, M.T.; Aliakbari, A.; Sahhafi, S.R. Evaluation of some physiological traits associated with improved drought tolerance in Iranian wheat. *Annu. Biol. Res.* **2012**, *3*, 1719–1725.

23. Yang, R.-C.; Jana, S.; Clarke, J.M. Phenotypic Diversity and Associations of Some Potentially Drought-responsive Characters in Durum Wheat. *Crop. Sci.* **1991**, *31*, 1484–1491. [CrossRef]

24. Sairam, R.K. Effect of moisture-stress on physiological activities of two contrasting wheat genotypes. *Indian J. Exp. Biol.* **1994**, *32*, 594.

25. David, M. Osmotic adjustment capacity and cuticular transpiration in several wheat cultivars cultivated in Algeria. *Rom. Agric. Res.* **2009**, *26*, 29–33.

26. Dong, B.; Shi, L.; Shi, C.; Qiao, Y.; Liu, M.; Zhang, Z. Grain yield and water use efficiency of two types of winter wheat cultivars under different water regimes. *Agric. Water Manag.* **2011**, *99*, 103–110. [CrossRef]

27. Åström, H.; Metsovuori, E.; Saarinen, T.; Lundell, R.; Hänninen, H. Morphological characteristics and photosynthetic capacity of Fragaria vesca L. winter and summer leaves. *Flora Morphol. Distrib. Funct. Ecol. Plants* **2015**, *215*, 33–39. [CrossRef]

28. Li, X.-W.; Jiang, J.; Zhang, L.-P.; Yu, Y.; Ye, Z.-W.; Wang, X.-M.; Zhou, J.-Y.; Chai, M.-L.; Zhang, H.-Q.; Arus, P.; et al. Identification of volatile and softening-related genes using digital gene expression profiles in melting peach. *Tree Genet. Genomes* **2015**, *11*. [CrossRef]

29. Bates, L.S.; Waldren, R.P.; Teare, I.D. Rapid determination of free proline for water-stress studies. *Plant Soil* **1973**, *39*, 205–207. [CrossRef]

30. Avestan, S.; Naseri, L.; Barker, A.V. Evaluation of nanosilicon dioxide and chitosan on tissue culture of apple under agar-induced osmotic stress. *J. Plant Nutr.* **2017**, *40*, 2797–2807. [CrossRef]

31. Chen, W.; Yao, X.; Cai, K.; Chen, J. Silicon alleviates drought stress of rice plants by improving plant water status, photosynthesis and mineral nutrient absorption. *Biol. Trace Elem. Res.* **2011**, *142*, 67–76. [CrossRef]

32. Yin, L.; Wang, S.; Liu, P.; Wang, W.; Cao, D.; Deng, X.; Zhang, S. Silicon-mediated changes in polyamine and 1-aminocyclopropane-1-carboxylic acid are involved in silicon-induced drought resistance in *Sorghum bicolor* L. *Plant Physiol. Biochem.* **2014**, *80*, 268–277. [CrossRef] [PubMed]

33. Gong, H.; Chen, K.; Chen, G.; Wang, S.; Zhang, C. Effects of Silicon on Growth of Wheat Under Drought. *J. Plant Nutr.* **2003**, *26*, 1055–1063. [CrossRef]

34. Hattori, T.; Sonobe, K.; Inanaga, S.; An, P.; Morita, S. Effects of Silicon on Photosynthesis of Young Cucumber Seedlings Under Osmotic Stress. *J. Plant Nutr.* **2008**, *31*, 1046–1058. [CrossRef]

35. Lu, C.; Zhang, C.; Wen, J.; Wu, G.; Tao, M. Research of the effect of nanometer materials on germination and growth enhancement of Glycine max and its mechanism. *Soybean Sci.* **2001**, *21*, 168–171.

36. Zheng, L.; Hong, F.; Lu, S.; Liu, C. Effect of Nano-TiO$_2$ on Strength of Naturally Aged Seeds and Growth of Spinach. *Boil. Trace Elem. Res.* **2005**, *104*, 083–092. [CrossRef]

37. Haghighi, M.; Pessarakli, M. Influence of silicon and nano-silicon on salinity tolerance of cherry tomatoes (*Solanum lycopersicum* L.) at early growth stage. *Sci. Hortic.* **2013**, *161*, 111–117. [CrossRef]

38. Nayyar, H.; Walia, D. Water Stress Induced Proline Accumulation in Contrasting Wheat Genotypes as Affected by Calcium and Abscisic Acid. *Boil. Plant.* **2003**, *46*, 275–279. [CrossRef]

39. Pei, Z.; Ming, D.F.; Liu, D.; Wan, G.L.; Geng, X.X.; Gong, H.J.; Zhou, W.J. Silicon improves the tolerance to water-deficit stress induced by polyethylene glycol in wheat (*Triticum aestivum* L.) seedlings. *J. Plant Growth Regul.* **2010**, *29*, 106–115. [CrossRef]

40. Lee, S.K.; Sohn, E.Y.; Hamayun, M.; Yoon, J.Y.; Lee, I.J. Effect of silicon on growth and salinity stress of soybean plant grown under hydroponic system. *Agrofor. Syst.* **2010**, *80*, 333–340. [CrossRef]

41. Jain, M.; Mathur, G.; Koul, S.; Sarin, N. Ameliorative effects of proline on salt stress-induced lipid peroxidation in cell lines of groundnut (*Arachis hypogaea* L.). *Plant Cell Rep.* **2001**, *20*, 463–468. [CrossRef]

42. Moussa, H.R. Influence of exogenous application of silicon on physiological response of salt-stressed maize (*Zea mays* L.). *Int. J. Agric. Biol.* **2006**, *8*, 293–297.

43. Surendar, K.K.S.K.K.; Devi, D.D.D.D.D.; Ravi, I.R.I.; Jeyakumar, P.J.P.; Kumar, S.R.K.S.R.; Velayudham, V.K. Impact of water deficit on epicuticular wax, proline and free amino acid content and yield of banana cultivars and hybrids. *Plant Gene Trait* **2013**, *4*. [CrossRef]

44. Zhu, J.-K. Salt and Drought Stress Signal Transduction in Plants. *Annu. Rev. Plant Boil.* **2002**, *53*, 247–273. [CrossRef]

45. Marafon, A.C.; Endres, L. Silicon: Fertilization and nutrition in higher plants. *Rev. Cienc. Amazon. J. Agric. Environ. Sci.* **2013**. [CrossRef]

46. Liang, Y.; Nikolic, M.; Bélanger, R.; Gong, H.; Song, A. *Silicon in Agriculture: From Theory to Practice*; Springer: New York, NY, USA, 2015; p. 235.

47. Romero-Aranda, M.R.; Jurado, O.; Cuartero, J. Silicon alleviates the deleterious salt effect on tomato plant growth by improving plant water status. *J. Plant Physiol.* **2006**, *163*, 847–855. [CrossRef]

48. Tuna, A.L.; Kaya, C.; Higgs, D.; Murillo-Amador, B.; Aydemir, S.; Girgin, A.R. Silicon improves salinity tolerance in wheat plants. *Environ. Exp. Bot.* **2008**, *62*, 10–16. [CrossRef]

49. Tahir, M.A.; Aziz, T.; Farooq, M.; Sarwar, G. Silicon-induced changes in growth, ionic composition, water relations, chlorophyll contents and membrane permeability in two salt-stressed wheat genotypes. *Arch. Agron. Soil Sci.* **2012**, *58*, 247–256. [CrossRef]

50. Zuccarini, P. Effects of silicon on photosynthesis, water relations and nutrient uptake of Phaseolus vulgaris under NaCl stress. *Boil. Plant.* **2008**, *52*, 157–160. [CrossRef]

51. Savant, N.K.; Korndörfer, G.H.; Datnoff, L.E.; Snyder, G.H. Silicon nutrition and sugarcane production: A review 1. *J. Plant Nutr.* **1999**, *22*, 1853–1903. [CrossRef]

52. Yaghubi, K.; Ghaderi, N.; Vafaee, Y.; Javadi, T. Potassium silicate alleviates deleterious effects of salinity on two strawberry cultivars grown under soilless pot culture. *Sci. Hortic.* **2016**, *213*, 87–95. [CrossRef]

53. Blum, A.; Shpiler, L.; Golan, G.; Mayer, J. Yield stability and canopy temperature of wheat genotypes under drought-stress. *Field Crop Res.* **1989**, *22*, 289–296. [CrossRef]

54. Ndiso, J.; Chemining'Wa, G.; Olubayo, F.; Saha, H. Effect of drought stress on canopy temperature, growth and yield performance of cowpea varieties. *Int. J. Plant Soil Sci.* **2016**, *9*, 1–12. [CrossRef]

55. Shepherd, T.; Griffiths, D.W. The effects of stress on plant cuticular waxes. *New Phytol.* **2006**, *171*, 469–499. [CrossRef] [PubMed]

56. Mohammed, S. The Role of Leaf Epicuticular Wax an Improved Adaptation to Moisture Deficit Environments in Wheat. Ph.D. Thesis, Texas A & M University, College Station, TX, USA, May 2014.

57. Keller, C.; Rizwan, M.; Davidian, J.C.; Pokrovsky, O.S.; Bovet, N.; Chaurand, P.; Meunier, J.D. Effect of silicon on wheat seedlings (*Triticum turgidum* L.) grown in hydroponics and exposed to 0 to 30 µM Cu. *Planta* **2015**, *241*, 847–860. [CrossRef]

58. Sahebi, M.; Hanafi, M.M.; Akmar, A.S.N.; Rafii, M.Y.; Azizi, P.; Tengoua, F.F.; Azwa, J.N.M.; Shabanimofrad, M. Importance of Silicon and Mechanisms of Biosilica Formation in Plants. *BioMed Int.* **2015**, *2015*, 1–16. [CrossRef] [PubMed]

59. Prychid, C.J.; Rudall, P.J.; Gregory, M. Systematics and Biology of Silica Bodies in Monocotyledons. *Bot. Rev.* **2003**, *69*, 377–440. [CrossRef]

Agronomy **2019**, *9*, 246

60. Fawe, A.; Menzies, J.G.; Chérif, M.; Bélanger, R.R. Silicon and disease resistance in dicotyledons. In *Studies Plant Science*; Elsevier: Amsterdam, The Netherlands, 2001; Volume 8, pp. 159–169.

61. Pozza, E.A.; Pozza, A.A.A.; Botelho, D.M.D.S. Silicon in plant disease control. *Rev. Ceres* **2015**, *62*, 323–331. [CrossRef]

62. Zhang, W.; Xie, Z.; Lang, D.; Cui, J.; Zhang, X. Beneficial effects of silicon on abiotic stress tolerance in legumes. *J. Plant Nutr.* **2017**, *40*, 2224–2236. [CrossRef]

63. Agarie, S.; Uchida, H.; Agata, W.; Kubota, F.; Kaufman, P.B. Effects of Silicon on Transpiration and Leaf Conductance in Rice Plants (*Oryza sativa* L.). *Plant Prod. Sci.* **1998**, *1*, 89–95. [CrossRef]

agronomy

MDPI

Article

Prohexadione-Calcium Application during Vegetative Growth Affects Growth of Mother Plants, Runners, and Runner Plants of Maehyang Strawberry

Hyeon Min Kim [1], Hye Ri Lee [1], Jae Hyeon Kang [2] and Seung Jae Hwang [1,2,3,4,5,*]

[1]   Division of Applied Life Science, Graduate School of Gyeongsang National University, Jinju 52828, Korea;
       s75364@daum.net (H.M.K.); dgpfl77@naver.com (H.R.L.)
[2]   Division of Crop Science, Graduate School of Gyeongsang National University, Jinju 52828, Korea;
       golbang225@naver.com
[3]   Department of Agricultural Plant Science, College of Agriculture & Life Science,
       Gyeongsang National University, Jinju 52828, Korea
[4]   Institute of Agriculture & Life Science, Gyeongsang National University, Jinju 52828, Korea
[5]   Research Institute of Life Science, Gyeongsang National University, Jinju 52828, Korea
*   Correspondence: hsj@gnu.ac.kr; Tel.: +82-010-6747-5485

Received: 30 January 2019; Accepted: 21 March 2019; Published: 25 March 2019

Abstract: Strawberry (*Fragaria* × *ananassa* Duch.) is an important horticultural crop that is vegetatively propagated using runner plants. To achieve massive production of runner plants, it is important to transfer the assimilation products of the mother plant to the runner plants, and not to the runner itself. Application of prohexadione–calcium (Pro–Ca), a plant growth retardant with few side effects, to strawberry is effective in inhibiting transport of assimilates to runners. This study aimed to determine the optimum application method and concentration of Pro–Ca on the growth characteristics of mother plants, runners, and runner plants for the propagation of strawberry in nurseries. Pro–Ca was applied at the rate of 0, 50, 100, 150, or 200 mg·L^{-1} (35 mL per plant) to plants via foliar spray or drenching under greenhouse conditions at 30 days after transplantation. Petiole lengths of mother plants were measured 15 weeks after treatment; growth was suppressed at the higher concentrations of Pro–Ca regardless of the application method. However, the crown diameter was not significantly affected by the application method or Pro–Ca concentration. The number of runners was 7.0 to 8.2, with no significant difference across treatments. Runner length was shorter at higher concentrations of Pro–Ca, especially in the 200 mg·L^{-1} drench treatment. However, fresh weight (FW) and dry weights (DW) of runners in the 50 mg·L^{-1} Pro–Ca drench treatments were higher than controls. Foliar spray and drench treatments were more effective for runner plant production than the control; a greater number of runner plants were produced with the 100 and 150 mg·L^{-1} Pro–Ca foliar spray treatment and the 50 and 100 mg·L^{-1} drench treatment. The FW and DW of the first runner plant was not significantly different in all treatments, but DW of the second runner plant, and FW and DW of the third runner plant were greatest in the 50 mg·L^{-1} Pro–Ca drench treatment. These results suggested that growth and production of runner plants of Maehyang strawberry were greatest under the 50 mg·L^{-1} Pro–Ca drench treatment.

Keywords: drench; foliar spray; *Fragaria* × *ananassa*; runner length

1. Introduction

Plant growth retardants (PGRs), like anti-gibberellins, have been used in agricultural industries for decades to improve the quality and quantity of horticultural crops [1,2]. PGRs such as daminozide, paclobutrazol, chlormequat chloride, uniconazole, and prohexadione–calcium (Pro–Ca) are used to control plant size and shape, specifically to reduce vegetative growth [3–6]. Among them, Pro–Ca has various advantages over other PGRs; it has negligible toxicological effects on mammals and a short persistence period in plants and soil [7,8]. In addition, Pro–Ca application delays senescence by lowering ethylene production within plants [9], and enhances resistance to disease and insects by inhibiting the biosynthesis of phenol [10]. Pro–Ca is a gibberellin (GA) biosynthesis inhibitor, which is the co-substrate for dioxygenases catalyzing hydroxylations involved in the late stages of GA biosynthesis. The main target of Pro–Ca seems to be 3β-hydroxylase, an enzyme that catalyzes primarily the conversion of inactive GA_{20}/GA_9 into highly active GA_1/GA_4 in either the early 13-hydroxylated pathway or the early non-13-hydroxylation pathway, respectively [11–13]. Reducing plant height is an important effect of PGR application, leading to increased quality and yield, and decreases in cost, space, and labor [13,14]. Pro–Ca has been shown to reduce and regulate the growth of crops such as petunia, impatiens, rice, chrysanthemum, pear, and various vegetables without the negative effects of decline in fruit quality and yield [8,13–16]. Most studies on the influences of Pro–Ca have focused on seed-propagated crops, but there are few reports on the application of Pro–Ca to vegetatively-propagated crops [16–18].

In fruit and vegetable crops, the quality of seedlings and other propagules are known to be very important for the quality and quantity of subsequent production. Accordingly, the quality of strawberry (*Fragaria* × *ananassa* Duch.) propagules has a direct influence on the yield and quality of fruit after transplanting; propagule quality is estimated to account for 80% of the whole crop cultivation quality [19]. Unlike fruit and vegetable crops such as tomato, cucumber, and watermelon, strawberry is distinctive in that it requires a lot of time and labor for the production of runner plants from vegetative organs. Strawberries are cultivated nurseries from the end of March to the beginning of September in the Republic of Korea. Various processes, such as transplanting of the strawberry mother plants, occurrence of runners and runner plants, fixation of runner plants, removal of runner plants from mother plants, and induction of flower bud differentiation require a period of five to six months [20,21]. Initiation of runner and runner plants occurs from May to June. Previous studies have focused on nutrient uptake, such as management of calcium fertilization [22], phosphorus [23], bicarbonate [24], and sulfur [25], of the strawberry mother plant. In addition, strawberry is known to be more sensitive to salinity than the other crops [26]. For that reason, previous studies have been conducted to determine the optimum electrical conductivity (EC) levels of nutrient solutions for mother and runner plants during the nursery period [27,28].

Although previous studies have reported production of large numbers of runners and runner plants, there are insufficient studies on runner length. During runner production, runners that are overly long are difficult to manage and require a considerable labor force within the restricted space of nurseries. Furthermore, production of runners and runner plants within high planting densities can reduce their quality and yield. Thus, there is a need for effective research to improve the quality and quantity of runner plants by shortening the runner without negatively effecting physiology.

In the present study, we hypothesize that the application method and concentration of Pro–Ca will improve the quantity and quality of strawberry runner plants by promoting their growth and development. To test our hypothesis, we investigated the growth of mother plants and propagation of runners and runner plants, and measured the biomass of the first, second, and third runner plants of the Maehyang cultivar strawberry for export in the Republic of Korea under greenhouse conditions, as well as confirmed the feasibility of practical application of the technology.

2. Materials and Methods

2.1. Plant Materials and Growth Conditions

The experiment was conducted in an even-span greenhouse (9 × 24 × 3 m) set up as a strawberry nursery with a hydroponic system and located at Gyeongsang National University in the Republic of Korea. Mother plants of strawberry (*Fragaria* × *ananassa* Duch. 'Maehyang') were planted at a density of four plants per pot using a strawberry cultivation container (61 × 27 × 18 cm, Hwaseong

Industrial Co. Ltd., Okcheon, Korea) filled with commercial strawberry-growing medium (BC2, BVB substrates Co. Ltd., De Lier, the Netherlands) on 20 March, 2018. During the cultivation period, the temperature of the even-span greenhouse was maintained at $26 \pm 5\,°C$ during the day and $16 \pm 5\,°C$ at night, $60 \pm 10\%$ relative humidity, and a natural photoperiod of 12–14 h. Well-rooted mother plants were fertilized using drip tape with Bas Van Buuren (BVB) strawberry solution from the Netherlands (in $mg \cdot L^{-1}$: $Ca(NO_3)_2 \cdot 4H_2O$ 613.0, KNO_3 187.0, KH_2PO_4 227.0, K_2SO_4 114.0, $MgSO_4 \cdot H_2O$ 275.0, NH_4NO_3 84.0, Fe–EDTA 10.60, H_3BO_3 0.31, $MnSO_4 \cdot 5H_2O$ 2.54, $ZnSO_4 \cdot 7H_2O$ 2.21, $CuSO_4 \cdot 5H_2O$ 0.16, and $Na_2MoO_4 \cdot 2H_2O$ 0.12, pH 5.8, and EC 1.5 $dS \cdot m^{-1}$). Chemical analysis of tap water revealed a composition of Ca^{2+} 0.40, Mg^{2+} 0.20, NH_4^+ 0.10, NO_3^- 0.10, HCO_3^- 0.71 $mmol \cdot L^{-1}$, pH 7.3, and EC 0.2 $dS \cdot m^{-1}$. During the cultivation period, 300 to 450 mL per culture pot was supplied two or three times (10 min per time), and the nutrient solution was adjusted to EC 1.5 $dS \cdot m^{-1}$ and pH 5.8. Prior to the treatment of strawberry mother plants with Pro–Ca, old leaves, axillary buds, and all runners were removed. Pesticides were applied every 7 days to control major diseases and insects, such as powdery mildew, anthracnose disease, *Bradysia agrestis*, aphids, and mites.

2.2. Application Methods and Concentration of Pro–Ca

Four different concentrations, 50, 100, 150, and 200 $mg \cdot L^{-1}$ of Pro–Ca (prohexadione–calcium, Sigma–Aldrich Co. Ltd., Saint Louis, MO, USA), were applied to plants via a foliar spray or drench. The Pro–Ca treatment for mother plants was applied one time under greenhouse conditions at 30 days after transplanting on 18 April, 2018. Foliar spray of Pro–Ca was applied using a hand sprayer, and the drench treatment of Pro–Ca was applied by slowly pouring it into the medium. The same volume of Pro–Ca solution (35 mL per plant) was applied in all treatments. Control plants were treated with tap water (35 mL per plant).

2.3. Measurements of Plant Growth Characteristics

Numbers of leaves, petiole length, soil plant analysis development (SPAD), number of runners, and runner length were measured each week after treatment with Pro–Ca for six weeks. The number of leaves and runners were counted by eye. Chlorophyll content was represented as the SPAD, which was measured using a portable chlorophyll meter (SPAD-502, Konica Minolta Inc., Tokyo, Japan). Growth parameters of mother plants, runners, and runner plants, such as petiole length, number of leaves, crown diameter, shoot fresh weights (FW), and dry weights (DW) of mother plants, leaf area, runner length, FW and DW of runners, number of runner plants, and FW and DW of the first, second, and third runner plants were measured at 15 weeks after treatment. The crown diameter was measured using a vernier caliper (CD-20CPX, Mitutoyo Co. Ltd., Kawasaki, Japan). Leaf area was measured using a leaf area meter (LI-3000, LI-COR Inc., Lincoln, NE, USA). The FW of the mother plant shoot, runner, and runner plants (first, second, and third) were measured using an electronic balance (EW220-3NM, Kern and Sohn GmbH, Balingen, Germany), and the DW of the mother plant shoot, runner, and the first, second, and third runner plant were measured the in a similar manner. Plant tissue was dried in an oven (Venticell-220, MMM Medcenter Einrichtungen GmbH, Planegg, Germany) at 70 °C for 72 h, and DW measured using an electronic balance.

2.4. Measurement of Chlorophyll Fluorescence (Fv/Fm)

For assessing photosystem II (PS II) performance, chlorophyll fluorescence measurements were taken from dark-adapted leaves of all treatments using a portable leaf fluorometer (FluorPen FP 100, Photon System Instruments, Drasov, Czech Republic). After dark adaptation for 30 min, chlorophyll fluorescence was measured on the upper surfaces of the leaves [29]. The minimum fluorescence (Fo) was obtained by measuring the light at 0.6 kHz and photosynthetic photon flux density (PPFD) below 0.1 $\mu mol \cdot m^{-2} \cdot s^{-1}$ using a red LED light. The maximum fluorescence (Fm) was measured by irradiating saturation light of 7000 $\mu mol \cdot m^{-2} \cdot s^{-1}$ at 20 kHz for 0.8 s. The variable/maximum fluorescence ratio (Fv/Fm) was calculated by the formula Fv/Fm = (Fm − Fo)/Fm [30]. Fv/Fm represents the maximum quantum yield of PS II photochemistry measured in the dark-adapted state. To measure the Fv/Fm, leaves of six mother plants were used for each treatment.

2.5. Statistical Analysis

The experimental treatments were randomized in a split-plot design, assigning the Pro–Ca application methods to the main plots and the Pro–Ca concentrations to the sub-plots. Each treatment included four plants and was repeated three times. The statistical analyses were performed using an SAS program (SAS 9.4, SAS Institute Inc., Cary, NC, USA). The experimental results were subjected to analysis of variance (ANOVA) and Tukey's tests. Graphing was performed with the SigmaPlot program (SigmaPlot 12.0, Systat Software Inc., San Jose, CA, USA).

3. Results and Discussion

3.1. Growth Characteristics of Mother Plants and Runners after Pro–Ca Treatment for Six Weeks

Figure 1 summarizes the growth characteristics of mother plants and runners of Maehyang strawberry treated with two application methods of four different Pro–Ca concentrations for six weeks. The petiole length of mother plants significantly decreased in the Pro–Ca treatment groups compared to control plants, regardless of the application method. Further, the higher concentration of Pro–Ca showed an inhibition effect on petiole length extension (Figure 1A). In previous studies, it has been shown that PGR treatment in *Spathiphyllum*, rice, chrysanthemum, cucumber, apple, and tomato was associated with suppression of plant stretchiness [8,14,31–34]. In the present study, the same results were obtained in the Pro–Ca treatment groups. Additionally, there was no negative effect on the development of new leaves in Pro–Ca treatment groups, except for treatments at four and five weeks with 200 mg·L^{-1} Pro–Ca applied as a drench (Figure 1B). Similarly, Reekie et al. [17] observed that the number of leaves was not affected by 62.5 mg·L^{-1} Pro–Ca applied as a foliar spray on the strawberry cultivars Sweet Charlie and Camarosa. Plant height was inhibited in tomato by Pro–Ca without affecting the leaf number during the seedling growth period [2]. These results are explained by the fact that GA regulates cell elongation rather than cell division. The SPAD was significantly higher in Pro–Ca treatment groups than in the control group during the four-week period after treatment, especially in the foliar spray treatment with 200 mg·L^{-1} Pro–Ca (Figure 1C). Generally, PGRs inhibited plant size and biomass while increasing chlorophyll content. This effect is presumably because PGRs reduce GA biosynthesis in the plant, furthermore, cell elongation was decreased, which is the main physiological function of Pro–Ca [35]. According to Yoon and Sagong [4], the leaf area of apple trees decreased, while the specific leaf area and chlorophyll content were increased by Pro–Ca treatment. In addition, Chinese cabbage treated with Pro–Ca at 400 mg·L^{-1} exhibited significantly increased chlorophyll content compared to non-treatment [1]. Appropriate inhibition of plant vegetative growth may increase light use efficiency under high plant density. Furthermore, the increase of SPAD could improve the photosynthetic rate of individual leaves. The Fv/Fm of plants grown under normal conditions is generally in the range 0.80 to 0.84, indicating the stress index and maximum quantum yield of PS II photochemistry [29,36] (Figure 1D). In the present study, we confirmed chemical stress was caused by Pro–Ca application, but that the range of normal growth conditions (0.80 to 0.84) occurred in all treatment groups, except for plants in the first week following treatment. On the contrary, Fv/Fm was 0.778 and 0.815 at weeks two and five, respectively, for the control group, which was lower than Pro–Ca treatment groups. Similarly, Ilias et al. [7] reported that Fv/Fm was not decreased by Pro–Ca treatment in the okra cultivars Psalidati and Clemson Spineless. These results suggest that Pro–Ca treatment does not impose a negative stress on mother plants of strawberry. Runner length was significantly shorted in the higher treatment concentrations of Pro–Ca; notably, runner length was significantly inhibited at 200 mg·L^{-1} Pro–Ca with the drench treatment (Figure 1E). Pro–Ca blocks the conversion of physiologically inactive GA$_{20}$/GA$_9$ into highly physiologically active GA$_1$/GA$_4$. Pro–Ca remains in the plant and medium for 3–4 weeks after treatment and inhibits vegetative growth. After that period, inhibition of endogenous GA biosynthesis by Pro–Ca decreases and vegetative growth resumes [37,38]. Runner length tended to increase after treatment with 50 and 100 mg·L^{-1} Pro–Ca as foliar spray and 50 mg·L^{-1} Pro–Ca as a drench 4–6 weeks after treatment. This result implied that the elongation of runners was due to the lower levels of residual Pro–Ca in the mother plant, which stimulated vegetative growth. The occurrence of runners did not differ significantly from the control treatment, except for 200 mg·L^{-1} Pro–Ca applied as a drench (Figure 1F).

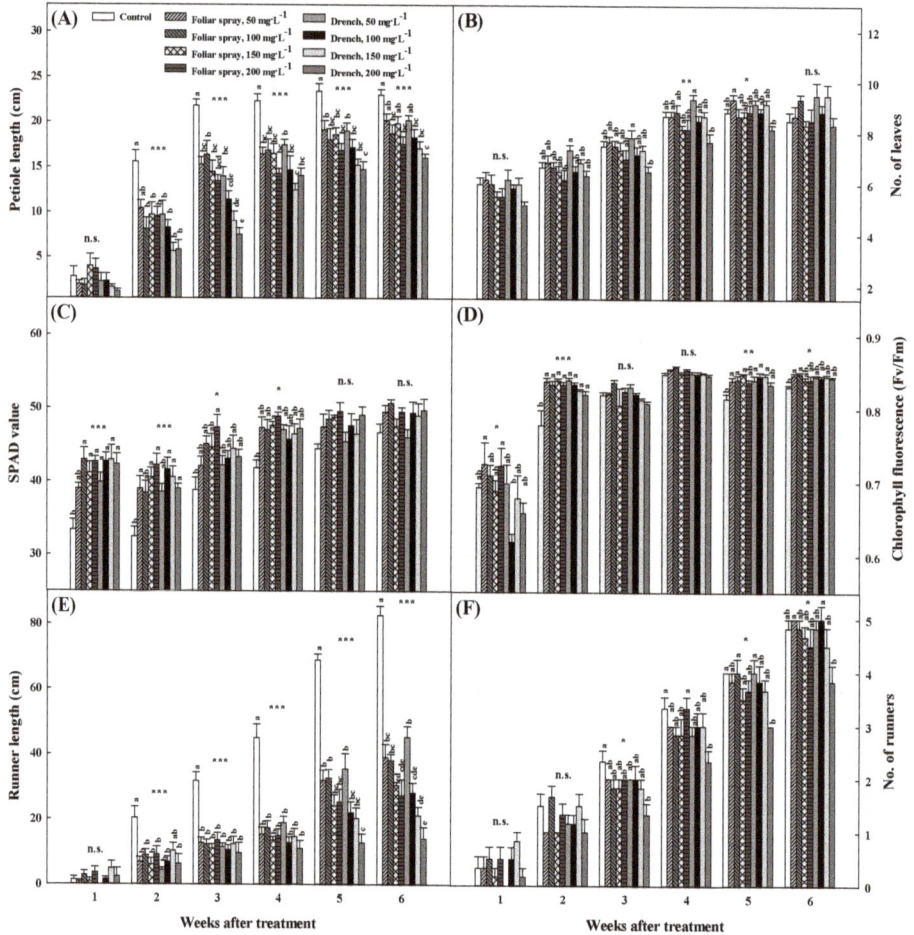

Figure 1. Petiole length (**A**), number of leaves (**B**), SPAD value (**C**), chlorophyll fluorescence (**D**), runner length (**E**), and number of runners (**F**) of strawberry cultivar Maehyang as affected by application method and concentration of prohexadione–calcium (Pro–Ca) at 1, 2, 3, 4, 5, and 6 weeks following treatment. Vertical bars represent standard deviation from the mean (*n* = 6). Different letters in the same column indicate significant differences based on Tukey's test ($p \leq 0.05$). n.s, *, **, *** no statistically significant difference or significant at $p \leq 0.05$, 0.01, and 0.001, respectively.

3.2. Growth Characteristics of Mother Plants at 15 Weeks after Treatment with Pro–Ca

The growth characteristics of the mother plant at 15 weeks following treatment with Pro–Ca, the method of application, and the concentration are shown in Table 1. The petiole length significantly inhibited drench application more than foliar spray. The combined FW and DW of leaves and petioles and leaf area were decreased more by the Pro–Ca drench application than the foliar spray, but there was no significant difference between the two application method treatment groups. In the treatment with PGRs, foliar spray was rapidly absorbed through the leaves and a larger amount of PGR was required. On the other hand, the effect of the drench application was slow but effective for a long time period even at low concentrations [39]. In the present study, the same amount (35 mL) of treatment was applied to all mother plants regardless of application method. Therefore, it is considered that the shorter petiole length was a function of drench application, but not foliar application. In terms of the Pro–Ca concentration, increasing the Pro–Ca concentration caused a reduction in the vegetative growth in the strawberry cultivar Maehyang. The petiole length was significantly inhibited at the 200 mg·L^{-1} concentration treatment. However, the crown diameter and the FW and DW of the crown were not significantly different between the treatment groups. In particular, the growth of the mother plant slowed at 200 mg·L^{-1} Pro–Ca applied as a drench, even when the residual period had passed. Reekie et al. [17] reported that in the strawberry cultivars Sweet Charlie and Camarosa, DW of leaf, stem, and root were similar to the control at 42 days after treatment with 62.5 mg·L^{-1} Pro–Ca as a foliar spray. However, in the present study, the lowest concentration of Pro–Ca (50 mg·L^{-1}) resulted in lower combined FW and DW of leaves and petioles, compared to the control. Generally, strawberry plants show different responses depending on the cultivar being treated [40,41], and these characteristics are controlled by genetic traits of the cultivars [42]. Barreto et al. [43] reported that concentrations of 200 and 400 mg·L^{-1} Pro–Ca markedly reduced vegetative growth indicators such as petiole length and leaf area in the strawberry cultivars Camarosa and Aromas. Thus, a concentration of 100 mg·L^{-1} Pro–Ca was suggested as the most appropriate concentration to apply to the cultivars Camarosa and Aromas. In the present study, however, the lower concentration 50 mg·L^{-1} Pro–Ca was sufficient to inhibit vegetative growth in the cultivar Maehyang. Therefore, it is likely that the effective concentration of Pro–Ca for vegetative growth inhibition will be different for each strawberry cultivar.

Table 1. Growth characteristics of the strawberry cultivar Maehyang mother plants as influenced by application method and concentration of Pro–Ca at 15 weeks after treatment.

Experiment Factor	Petiole Length (cm)	Crown Diameter (mm)	Fresh Weight (g/plant)		Dry Weight (g/plant)		Leaf Area (cm^2/plant)
			Leaves + Petioles	Crown	Leaves + Petioles	Crown	
			Application method				
Foliar spray	18.6	16.5	72.6	6.9	17.2	1.4	1561.1
Drench	16.4	16.8	67.3	7.2	15.6	1.5	1449.3
	*	n.s.	n.s.	n.s.	n.s.	n.s.	n.s.
			Concentration (mg·L^{-1})				
Control (0)	23.9 [a]	17.0	87.2 [a]	6.8	26.9 [a]	1.4	1867.3 [a]
50	19.3 [b]	16.3	75.5 [ab]	7.4	18.0 [b]	1.5	1538.7 [b]
100	18.0 [bc]	16.4	71.5 [b]	7.3	16.3 [b]	1.5	1541.0 [b]
150	17.4 [bc]	16.8	66.1 [b]	6.3	15.6 [b]	1.3	1557.7 [b]
200	15.5 [c]	17.1	66.8 [b]	7.2	15.6 [b]	1.5	1383.4 [b]
		n.s.		n.s.		n.s.	

Within each column, * significant difference at $p \leq 0.05$; n.s. no statistically significant difference; means followed by different letters are significantly different according to the Tukey's test at $p \leq 0.05$.

3.3. Growth Characteristics of Runners and Runner Plants at 15 Weeks after Treatment with Pro–Ca

There was no significant difference in the number of runners irrespective of application method and concentration of Pro–Ca at 15 weeks after treatment (Figure 2A). Foliar spray and drenching of Pro–Ca were more effective for runner plant production than the control. The greatest number of runner plants were produced by applications of 100 and 150 mg·L^{-1} Pro–Ca by foliar spray, and 50 and 100 mg·L^{-1} Pro–Ca by drenching (Figure 2B). The number of runner plants per mother plant was the lowest (10.6) after application of 200 mg·L^{-1} as a drench, which was lower than the control. In previous studies, foliar spray application of GA produced large numbers of runners and runner plants of strawberries in the nursery period [42,44]. In the present study, however, Pro–Ca, an anti-gibberellin was more effective for the production of runner plants using both application methods and at concentration of 50 to 150 mg·L^{-1}, compared to the control. These results implied that the assimilation products used for the growth of mother plants were more effectively translocated for the development of runners and runner plants after application of Pro–Ca.

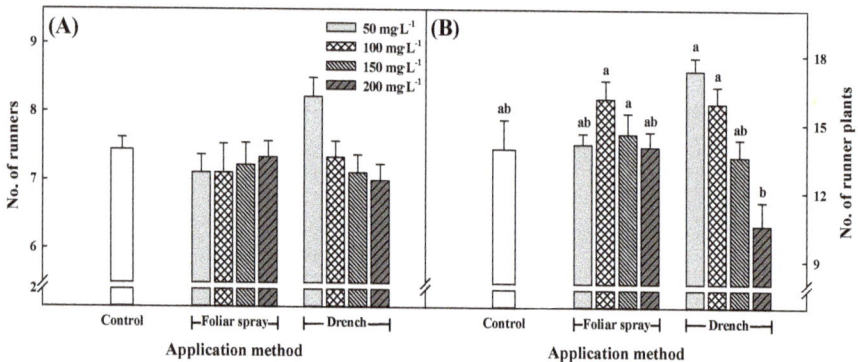

Figure 2. Number of runners (**A**) and runner plants (**B**) of the strawberry cultivar Maehyang as affected by application method and concentration of Pro–Ca at 15 weeks after treatment. Vertical bars represent the standard deviation of the mean ($n = 9$). Different letters in the same column indicate significant differences based on Tukey's test ($p \leq 0.05$).

The FW and DW of runners were similar to those of runner plants (Figure 3A,B). As the concentration of Pro–Ca increased, the FW and DW of runners decreased, and especially, negative correlations were observed from plants treated by drench application. In tomato, plant height was reduced more effectively by Pro–Ca applied as a drench application than as a foliar spray [2]. The results of this study also showed that reducing the FW and DW of the runners occurred more effectively as the concentration increased in the drench application rather than the foliar spray. According to Savini et al. [45], the runner acts as a transporter to translocate assimilates, nutrient elements, and water from the mother plant to the runner plant. Therefore, heavier biomass of the runners has a positive effect on plant-to-plant communication. Consequently, the increased FW and DW of the runner was a positive achievement of the application of 50 mg·L^{-1} Pro–Ca as a drench.

Total runner length and comparison of runner lengths are shown in Figure 4A,B. The total runner length was shorter in all Pro–Ca treatment groups than in the control except for 50 mg·L^{-1} applied as a drench. However, the comparison of runner length from the mother plant to the first runner plant, from the first runner plant to the second runner plant, and from the second runner plant to the third runner plant showed that they were shorter after application of 50 mg·L^{-1} Pro–Ca as a drench than in the control. Similar results were obtained by Hytönen et al. [16], who obtained reduced elongation of runners by application of 50 mg·L^{-1} Pro–Ca with foliar spray compared to the non-treatment in the strawberry cultivar Korona. The total runner length and comparison of runner length was shortest after treatment with 200 mg·L^{-1} Pro–Ca as a drench. The runner length has a great influence on the determination of bed height in strawberry high bench type culture, and furthermore, the runner length tends to be inversely proportional to the runner diameter [46]. Therefore, it is considered that reducing runner length helps to produce higher quality runners and runner plants.

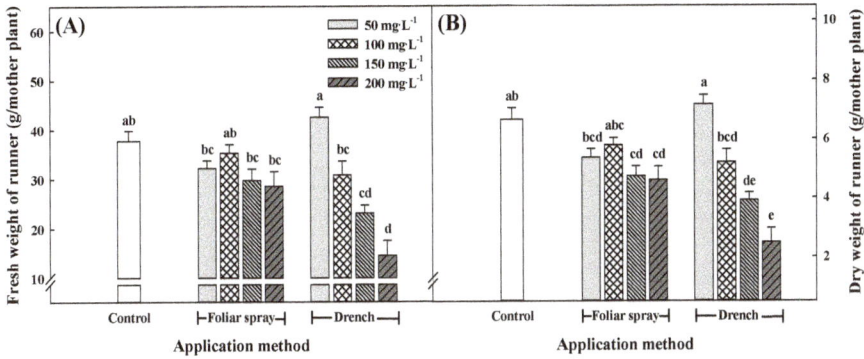

Figure 3. Fresh weight of runners (**A**) and dry weight of runners (**B**) of the strawberry cultivar Maehyang as affected by application method and concentration of Pro–Ca at 15 weeks after treatment. Vertical bars represent the standard deviation of the mean (*n* = 9). Different letters in the same column indicate significant differences based on Tukey's test (*p* ≤ 0.05).

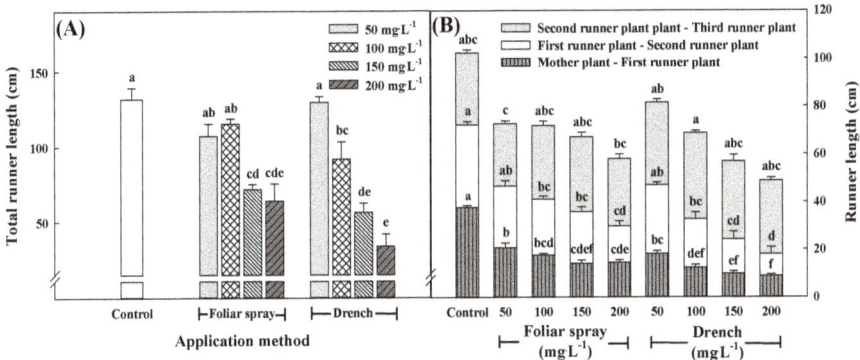

Figure 4. Total runner length (**A**) and comparison of runner length (**B**) of the strawberry cultivar Maehyang as affected by application method and concentration of Pro–Ca at 15 weeks after treatment. Vertical bars represent the standard deviation of the mean (*n* = 9). Different letters in the same column indicate significant differences based on Tukey's test (*p* ≤ 0.05).

3.4. Growth Characteristics of the First, Second, and Third Runner Plants at 15 Weeks after Treatment with Pro–Ca

The growth characteristics of runner plants as affected by application method and concentration of Pro–Ca at 15 weeks after treatment are shown in Table 2. The FW and DW of the first runner plant showed no significant difference under all treatments. The FW of the second and third runner plants showed a tendency to be heavier in the foliar spray than the drench application, and the 50 mg·L^{-1} Pro–Ca concentration resulted in higher FW of the second and third runner plant than the control. In addition, DW of the second and third runner plants were heavier after the 50 mg·L^{-1} Pro–Ca application than the other concentrations of treatments. According to Reekie et al. [17], the net photosynthetic rate of mother and runner plants of the strawberry cultivars Sweet Charlie and Camarosa was increased by application of 62.5 mg·L^{-1} Pro–Ca as a foliar spray. Similarly, Sabatini et al. [47] reported that Pro–Ca positively affected leaf mass area and chlorophyll content in apple and pear trees because net photosynthesis was increased after Pro–Ca application. Also, strawberry plants treated with Pro–Ca exhibited increased total DW, relative growth rate, and unit leaf rate during the nursery period [18]. Moreover, strawberry cultivar Camarosa runner plants treated with 100 mg·L^{-1} Pro–Ca as a foliar spray exhibited a photosynthetic rate increase of 23% compared to non-treated controls [48]. This physiological phenomenon caused by Pro–Ca application occurs as a result of light energy being converted into chemical energy (ATP and NADPH) that is used to reduce atmospheric CO_2 to carbohydrates through the Calvin cycle during photosynthesis [49]. Thus, in the present study, we propose that the net photosynthetic rate was increased by Pro–Ca treatment. In addition, it is considered that the photosynthetic products of the mother plant were more effectively translocated to the runner plant due to the formation of the high quality runner after application of 50 mg·L^{-1} Pro–Ca (Figure 3). In previous studies, Pro–Ca treatment of strawberries focused mainly on the concentration and number of foliar spray application [16–18,43]. Foliar spray applications are most commonly used in PGRs, but can result in non-uniform plant size if careful attention to technique is not used [50]. On the other hand, soil or growth medium drench applications give more uniform results and increased product efficiency at lower concentrations compared to foliar spray application [51]. In general, PGRs are applied directly on the soil or growth medium as a drench, or as a foliar spray, however, it is known that variation in responses occurs among species and cultivars [52]. Here, we determined the best concentration and method of application of Pro–Ca to the strawberry cultivar Maehyang for optimum propagation of runners and runner plants with higher biomass is 50 mg·L^{-1} Pro–Ca applied as a soil or growth medium drench.

Table 2. Growth characteristics of strawberry cultivar Maehyang runner plants as affected by application method and concentration of Pro–Ca at 15 weeks after treatment.

Experiment Factor	Fresh Weight (g/plant)			Dry Weight (g/plant)		
	First Runner Plant	Second Runner Plant	Third Runner Plant	First Runner Plant	Second Runner Plant	Third Runner Plant
Application method						
Foliar spray	19.2	13.4	4.5	3.9	2.7	0.78
Drench	17.2	11.2	3.3	3.7	2.4	0.67
	n.s.	*	*	n.s.	n.s.	n.s.
Concentration (mg·L^{-1})						
Control (0)	17.5	8.7 [b]	3.5 [b]	4.0	1.9 [b]	0.72 [abc]
50	18.5	13.3 [a]	5.3 [a]	3.9	2.9 [a]	1.02 [a]
100	19.3	13.7 [a]	4.4 [ab]	4.1	2.8 [a]	0.82 [ab]
150	17.3	10.4 [ab]	2.9 [b]	3.5	2.1 [b]	0.50 [c]
200	17.7	11.7 [ab]	3.1 [b]	3.6	2.3 [ab]	0.57 [bc]
	n.s.			n.s.		

Within each column, * significant difference at $p \leq 0.05$; n.s. no statistically significant difference; means followed by different letters are significantly different according to the Tukey's test at $p \leq 0.05$.

4. Conclusions

The present study revealed that inhibiting the growth of mother plants using drench application of Pro–Ca caused more sensitive responses compared to a foliar spray. Runners had heavier biomass under low concentrations of Pro–Ca (50 mg·L^{-1}) applied as a drench. In addition, application of Pro–Ca (50 mg·L^{-1}) applied as a drench increased the initiation of runner plants and stimulated higher biomass, as measured by DW, of the second runner plant, and the FW and DW of the third runner plant. Overall, the results suggest that 50 mg·L^{-1} Pro–Ca applied as a soil drench is the most suitable way to promote the quality and quantity of strawberry cultivar Maehyang. This knowledge is expected to be beneficial for the practical management of mother plants, runners, and the propagation of runner plants during the nursery period.

Author Contributions: Conceptualization, S.J.H.; methodology, S.J.H. and H.M.K.; formal analysis, H.M.K., H.R.L., and J.H.K.; resources, S.J.H.; data curation, H.M.K.; writing—original draft preparation, H.M.K.; writing—review and editing, S.J.H.; project administration, S.J.H.; funding acquisition, S.J.H., H.M.K., H.R.L., and J.H.K.

Funding: This research was funded by the Agrobio-Industry Technology Development Program; Ministry of Food, Agriculture, Forestry, and Fisheries; Republic of Korea (Project No. 315004-5).

Acknowledgments: This research was supported by the Agrobio-Industry Technology Development Program; Ministry of Food, Agriculture, Forestry, and Fisheries; Republic of Korea (Project No. 315004-5).

Conflicts of Interest: The authors declare no conflict of interest.

References

1. Kang, S.M.; Kim, J.T.; Hamayun, M.; Hwang, I.C.; Khan, A.L.; Kim, Y.H.; Lee, J.H.; Lee, I.J. Influence of prohexadione-calcium on growth and gibberellins content of Chinese cabbage grown in alpine region of South Korea. *Sci. Hortic.* **2010**, *125*, 88–92. [CrossRef]
2. Altintas, S. Effects of chlormequat chloride and different rates of prohexadione-calcium on seedling growth, flowering, fruit development and yield of tomato. *Afr. J. Biotechnol.* **2011**, *10*, 17160–17169.
3. Latimer, J.G. Growth retardants affect landscape performance of *Zinnia, Impatiens,* and marigold. *HortScience* **1991**, *26*, 557–560. [CrossRef]
4. Yoon, T.M.; Sagong, D.H. Growth control of 'Fuji' apple trees by use of prohexadione-calcium. *Korean J. Hortic. Sci. Technol.* **2005**, *23*, 269–274.
5. Hwang, I.C.; Lee, I.J.; Cho, T.K.; Kim, J.T.; Yoon, C.S. Development of labor-saving plant growth regulator, prohexadione-calcium, for growth inhibition. *Korean J. Weed Sci.* **2009**, *29*, 1–8.
6. Kim, H.C.; Cho, Y.H.; Ku, Y.G.; Hwang, S.J.; Bae, J.H. Growth characteristics of grafted tomato seedlings following treatment with various concentrations of diniconazole during the summer growth season. *Korean J. Hortic. Sci. Technol.* **2016**, *34*, 249–256.
7. Ilias, I.; Ouzounidou, G.; Giannakoula, A.; Papadopoulou, P. Effects of gibberellic acid and prohexadione-calcium on growth, chlorophyll fluorescence and quality of okra plant. *Biol. Plant* **2007**, *51*, 575–578. [CrossRef]
8. Kim, H.Y.; Lee, I.J.; Hamayun, M.; Kim, J.T.; Won, J.G.; Hwang, I.C.; Kim, K.U. Effect of prohexadione-calcium on growth components and endogenous gibberellins contents of rice (*Oryza sativa* L.). *J. Agron. Crop Sci.* **2007**, *193*, 445–451. [CrossRef]
9. Medjdoub, R.; Val, J.; Blanco, A. Prohexadione-Ca inhibits vegetative growth of 'Smoothee Golden Delicious' apple trees. *Sci. Hortic.* **2004**, *101*, 243–253. [CrossRef]
10. Byers, R.E.; Yoder, K.S. Prohexadione-calcium inhibits apple, but not peach, tree growth, but has little influence on apple fruit thinning or quality. *HortScience* **1999**, *34*, 1205–1209. [CrossRef]
11. Brown, R.G.S.; Kawaide, H.; Yang, Y.Y.; Rademacher, W.; Kamiya, Y. Daminozide and prohexadione have similar modes of action as inhibitors of the late stages of gibberellin metabolism. *Physiol. Plant* **1997**, *101*, 309–313. [CrossRef]
12. Evans, J.R.; Evans, R.R.; Regusci, C.L.; Rademacher, W. Mode of action, metabolism, and uptake of BAS 125W, prohexadione-calcium. *HortScience* **1999**, *34*, 1200–1201. [CrossRef]
13. Ito, A.; Sakamoto, D.; Itai, A.; Nishijima, T.; Oyama-Okudo, N.; Nakamura, Y.; Moriguchi, T.; Nakajima, I. Effects of GA$_{3+4}$ and GA$_{4+7}$ application either alone or combined with prohexadione-ca on fruit development of Japanese pear 'Kosui'. *Hortic. J.* **2016**, *85*, 201–208. [CrossRef]
14. Kim, Y.H.; Khan, A.L.; Hamayun, M.; Kim, J.T.; Lee, J.H.; Hwang, I.C.; Yoon, C.S.; Lee, I.J. Effects of prohexadione calcium on growth and gibberellins contents of *Chrysanthemum morifolium* R. cv Monalisa White. *Sci. Hortic.* **2010**, *123*, 423–427. [CrossRef]
15. Ilias, I.; Rajapakse, N. Prohexadione-calcium affects growth and flowering of petunia and impatiens grown under photoselective films. *Sci. Hortic.* **2005**, *106*, 190–202. [CrossRef]
16. Hytönen, T.; Elomaa, P.; Moritz, T.; Junttila, O. Gibberellin mediates daylength-controlled differentiation of vegetative meristems in strawberry (*Fragaria × ananassa* Duch). *BMC Plant Biol.* **2009**, *9*, 18. [CrossRef] [PubMed]
17. Reekie, J.Y.; Hicklenton, P.R.; Struik, P.C. Prohexadione-calcium modifies growth and increases photosynthesis in strawberry nursery plants. *Can. J. Plant Sci.* **2005**, *85*, 671–677. [CrossRef]
18. Reekie, J.Y.; Struik, P.C.; Hicklenton, P.R.; Duval, J.R. Dry matter partitioning in a nursery and a plasticulture fruit field of strawberry cultivars 'Sweet Charlie' and 'Camarosa' as affected by prohexadione-calcium and partial leaf removal. *Eur. J. Hortic. Sci.* **2007**, *72*, 122–129.

19. Jun, H.J.; Jeon, E.H.; Kang, S.I.; Bae, G.H. Optimum nutrient solution strength for Korean strawberry cultivar 'Daewang' during seedling period. *Korean J. Hortic. Sci. Technol.* **2014**, *32*, 812–818. [CrossRef]
20. Na, Y.W.; Jeong, H.J.; Lee, S.Y.; Choi, H.G.; Kim, S.H.; Rho, I.R. Chlorophyll fluorescence as a diagnostic tool for abiotic stress tolerance in wild and cultivated strawberry species. *Hortic. Environ. Biotechnol.* **2014**, *55*, 280–286. [CrossRef]
21. Park, G.S.; Choi, J.M. Medium depths and fixation dates of 'Seolhyang' strawberry runner plantlets in nursery field influence the seedling quality and early growth after transplanting. *Korean J. Hortic. Sci. Technol.* **2015**, *33*, 518–524. [CrossRef]
22. Choi, J.M.; Nam, M.H.; Lee, H.S.; Kim, D.Y.; Yoon, M.K.; Ko, K.D. Influence of Ca fertilization on the growth and appearance of physiological disorders in mother plants and occurrence of daughter plants in propagation of 'Seolhyang' strawberry through soil cultivation. *Korean J. Hortic. Sci. Technol.* **2012**, *30*, 657–663. [CrossRef]
23. Choi, J.M.; Latigui, A.; Lee, C.W. Visual symptom and tissue nutrient contents in dry matter and petiole sap for diagnostic criteria of phosphorus nutrition for 'Seolhyang' strawberry cultivation. *Hortic. Environ. Biotechnol.* **2013**, *54*, 52–57. [CrossRef]
24. Lee, H.S.; Choi, J.M.; Kim, T.I.; Kim, H.S.; Lee, I.H. Influence of bicarbonate concentrations in nutrient solution on the growth, occurrence of daughter plants and nutrient uptake in vegetative propagation of 'Seolhyang' strawberry. *Korean J. Hortic. Sci. Technol.* **2014**, *32*, 149–156. [CrossRef]
25. Lee, H.S.; Park, I.S.; Choi, J.M. Influence of sulfur concentration on bicarbonate injury reduction during vegetative growth in 'Seolhyang' strawberry. *Hortic. Sci. Technol.* **2018**, *36*, 362–369.
26. Caruso, G.; Villari, G.; Melchionna, G.; Conti, S. Effects of cultural cycles and nutrient solutions on plant growth, yield and fruit quality of alpine strawberry (*Fragaria vesca* L.) grown in hydroponics. *Sci. Hortic.* **2011**, *129*, 479–485. [CrossRef]
27. Kim, H.M.; Kim, H.M.; Jeong, H.W.; Lee, H.R.; Jeong, B.R.; Kang, N.J.; Hwang, S.J. Growth of mother plants and occurrence of daughter plants of 'Maehyang' strawberry as affected by different EC levels of nutrient solution during nursery period. *Protected Hortic. Plant Fac.* **2018**, *27*, 185–190. [CrossRef]
28. Narváez-Ortiz, W.A.; Lieth, J.H.; Grattan, S.R.; Benavides-Mendoza, A.; Evans, R.Y.; Preciado-Rangel, P.; Valenzuela-García, J.R.; Gonzalez-Fuentes, J.A. Implications of physiological integration of stolon interconnected plants for salinity management in soilless strawberry production. *Sci. Hortic.* **2018**, *241*, 124–130. [CrossRef]
29. Maxwell, K.; Johnson, G.N. Chlorophyll fluorescence—A practical guide. *J. Exp. Bot.* **2000**, *51*, 659–668. [CrossRef]
30. Genty, B.; Briantais, J.M.; Baker, N.R. The relationship between the quantum yield of photosynthetic electron transport and quenching of chlorophyll fluorescence. *Biochim. Biophys. Acta* **1989**, *990*, 87–92. [CrossRef]
31. Won, E.J.; Jeong, B.R. Effect of plant growth retardants on the growth characteristics of potted *Spathiphyllum* in ebb and flow system. *Korean J. Hortic. Sci. Technol.* **2007**, *25*, 443–450.
32. Sun, E.S.; Kang, H.M.; Kim, Y.S.; Kim, I.S. Effects of seed soaking treatment of diniconazol on the inhibition of stretching of tomato and cucumber seedlings. *J. Bio-Environ. Control* **2007**, *19*, 55–62.
33. Sagong, D.H.; Song, Y.Y.; Park, M.Y.; Kweon, H.J.; Kim, M.J.; Yoon, T.M. Photosynthesis, shoot growth and fruit quality in 'Fuji'/M.9 mature apple trees in response to prohexadione-calcium treatments. *Korean J. Hortic. Sci. Technol.* **2014**, *32*, 762–770. [CrossRef]
34. Agehara, S.; Leskovar, D.I. Growth suppression by exogenous abscisic acid and uniconazole for prolonged marketability of tomato transplants in commercial conditions. *HortScience* **2017**, *52*, 606–611. [CrossRef]
35. Kofidis, G.; Giannakoula, A.; Ilias, I.F. Growth, anatomy and chlorophyll fluorescence of coriander plants (*Coriandrum sativum* L.) treated with prohexadione-calcium and daminozide. *Acta Biol. Crac. Ser. Bot.* **2008**, *50*, 55–62.
36. Baker, N.R.; Rosenqvist, E. Applications of chlorophyll fluorescence can improve crop production strategies: An examination of future possibilities. *J. Exp. Bot.* **2004**, *55*, 1607–1621. [CrossRef]
37. Evans, R.R.; Evans, J.R.; Rademacher, W. Prohexadione-calcium for suppression of vegetative growth in eastern apples. *Acta Hortic.* **1997**, *451*, 663–666. [CrossRef]
38. Schupp, J.R.; Robinson, T.L.; Cowgill, W.P.J.; Compton, J.M. Effect of water conditioner and surfactants on vegetative growth control and fruit cracking of 'Empire' apple caused by prohexadione-calcium. *HortScience* **2003**, *36*, 1205–1209. [CrossRef]

39. Lee, M.Y. Suppression of Stretchiness in Pot Kalanchoe by Various Applications of Plant Growth Retardants. Master's Thesis, Gyeongsang National University, Jinju, Korea, 2003.

40. Kender, W.J.; Carpenter, S.; Braun, J.W. Runner formation in ever-bearing strawberry as influenced by growth-promoting and inhibiting substances. *Ann. Bot.* **1971**, *35*, 1045–1052. [CrossRef]

41. Singh, J.P.; Randhawa, G.S.; Jain, N.L. Response of strawberry to gibberellic acid. *Indian J. Hortic. Sci.* **1960**, *17*, 21–30.

42. Momenpour, A.; Taghavi, T.S.; Manochehr, S. Effects of benzyladenine and gibberellin on runner production and some vegetative traits of three strawberry cultivars. *Afr. J. Agr. Res.* **2011**, *6*, 4357–4361.

43. Barreto, C.F.; Ferreira, L.V.; Costa, S.I.; Schiavon, A.V.; Becker, T.B.; Vignolo, G.K.; Antunes, L.E.C. Concentration and periods of application of prohexadione-calcium in the growth of strawberry seedlings. *Semina: Ciências Agrárias* **2018**, *39*, 1937–1944. [CrossRef]

44. Pipattanawong, N.; Fujishige, N.; Yamane, K.; Ijiro, Y.; Ogata, R. Effects of growth regulators and fertilizer on runner production, flowering, and growth in day-neutral strawberries. *Jpn. J. Trop. Agric.* **1996**, *40*, 101–105.

45. Savini, G.; Giorgi, V.; Scarano, E.; Neri, D. Strawberry plant relationship through the stolon. *Physiol. Plant* **2008**, *134*, 421–429. [CrossRef]

46. Kim, T.I.; Kim, W.S.; Choi, J.H.; Jang, W.S.; Seo, K.S. Comparison of runner production and growth characteristics among strawberry cultivars. *Korean J. Hortic. Sci. Technol.* **1999**, *17*, 111–114.

47. Sabatini, E.; Noferini, M.; Fiori, G.; Grappadelli, L.C.; Costa, G. Prohexadione-ca positively affects gas exchanges and chlorophyll content of apple and pear trees. *Eur. J. Hortic. Sci.* **2003**, *68*, 123–128.

48. Pereira, I.D.S.; Goncalves, M.A.; Picolotto, L.; Vignolo, G.K.; Antunes, L.E.C. Prohexadione-calcium growth control of 'Camarosa' strawberry seedlings cultivated in commercial substrate. *Amazon. J. Agric. Environ. Sci.* **2016**, *59*, 93–98.

49. Hofius, D.; Börnke, F.A.J. Photosynthesis, carbohydrate metabolism and source-sink relations. In *Potato Biology and Biotechnology*; Elsevier Science B.V.: Amsterdam, The Netherlands, 2007; pp. 257–285.

50. Barrett, J.E.; Bartuska, C.A.; Nell, T.A. Application techniques alter uniconazole efficacy on chrysanthemums. *HortScience* **1994**, *29*, 893–895. [CrossRef]

51. Hwang, S.J.; Lee, M.Y.; Park, Y.H.; Sivanesan, I.; Jeong, B.R. Suppression of stem growth in pot kalanchoe 'Gold Strike' by recycled subirrigational supply of plant growth retardants. *Afr. J. Biotechnol.* **2008**, *7*, 1487–1493.

52. Shin, W.G.; Hwang, S.J.; Sivanesan, I.; Jeong, B.R. Height suppression of tomato plug seedlings by an environment friendly seed treatment of plant growth retardants. *Afr. J. Biotechnol.* **2009**, *8*, 4100–4107.

agronomy

MDPI

Article

Assessment of Ultrasound Assisted Extraction as an Alternative Method for the Extraction of Anthocyanins and Total Phenolic Compounds from Maqui Berries (*Aristotelia chilensis* (Mol.) Stuntz)

Mercedes Vázquez-Espinosa, Ana V. González de Peredo, Marta Ferreiro-González, Ceferino Carrera, Miguel Palma, Gerardo F. Barbero * and Estrella Espada-Bellido

Department of Analytical Chemistry, Faculty of Sciences, University of Cadiz, Agrifood Campus of International Excellence (ceiA3), IVAGRO, 11510 Puerto Real, Cadiz, Spain; mercedes.vazquez@uca.es (M.V.-E.); ana.velascogope@uca.es (A.V.G.d.P.); marta.ferreiro@uca.es (M.F.-G.); ceferino.carrera@uca.es (C.C.); miguel.palma@uca.es (M.P.); estrella.espada@uca.es (E.E.-B.)
* Correspondence: gerardo.fernandez@uca.es; Tel.: +34-956-016355; Fax: +34-956-016460

Received: 15 February 2019; Accepted: 18 March 2019; Published: 21 March 2019

Abstract: Research interest regarding maqui (*Aristotelia chilensis*) has increased over the last years due to its potential health benefits as one of the most antioxidant-rich berries. Ultrasound-assisted extraction (UAE) is an advanced green, fast, and ecological extraction technique for the production of high quality extracts from natural products, so it has been proposed in this work as an ideal alternative extraction technique for obtaining extracts of high bioactivity from maqui berries. In order to determine the optimal conditions, the extraction variables (percentage of methanol, pH, temperature, ratio "sample mass/volume of solvent", amplitude, and cycle) were analyzed by a Box-Behnken design, in conjunction with the response surface method. The statistical analysis revealed that the temperature and the percentage of methanol were the most influential variables on the extraction of the total phenolic compounds and total anthocyanins, respectively. The optimal extraction time was determined at 15 min for total phenolic compounds, while it was only 5 min for anthocyanins. The developed methods showed a high precision level with a coefficient of variation of less than 5%. Finally, the new methods were successfully applied to several real samples. Subsequently, the results were compared to those that were obtained in previous experiments by means of microwave assisted extraction (MAE). Similar extraction yields were obtained for phenolic compounds under optimized conditions. However, UAE proved to be slightly more efficient than MAE in the extraction of anthocyanins.

Keywords: anthocyanins; *Aristotelia chilensis* (Mol.) Stuntz; maqui berry; food analysis; phenolic compounds; superfruit; ultrasound assisted extraction

1. Introduction

In recent years, there has been growing attention for the consumption of food rich in bioactive compounds that are associated to an improvement of health. Small berries, amongst other types of food, represent a diverse group that includes a number of rather small size, perishable, red, blue, and purple fruits, which are highly valued for their intense colour, delicate texture, and unique flavour [1]. Nowadays, there has been a considerable interest in finding natural antioxidants from plant materials to replace the synthetic ones. Berries are a rich source of bioactive compounds, which contribute to their antioxidant activity and different biological functions, so they can prevent diseases and health disorders [2]. Therefore, the interest on them and their analysis has grown enormously. Maqui (*Aristotelia chilensis* (Mol.) Stuntz), which is a shrub with reddish stems and evergreen leaves, is native to South America, and it grows in dense thickets forming wild populations, called "macales". It is a dioic berry from the Elaeocarpaceae family and it can grow up to 3–5 m tall. Between December and January, it produces small edible berries of a purple/black colour [3,4].

These berries are extremely rich in phenolic compounds and mainly anthocyanins, which are antioxidant substances that can remove free radicals and by its oxidation inhibit chain reactions. These compounds give maqui berries their intense blackish colour making into one of the berries with the most intense antioxidant properties known so far [5,6]. Special attention has recently been paid to these biological compounds, not only for their use as natural colorants, but also for their use in food and pharmaceutical industries for their disease preventive properties [7,8], as well as a food supplement or functional food product [9,10]. Its consumption can protect you against some chronic diseases, such as cardiovascular disorders, since, thanks to their antioxidants content, can prevent cholesterol from oxidizing in blood. They can also contribute to obesity control by accelerating metabolism and fat burning. Anti-inflammatory, anticarcinogenic, and antidiabetic properties, as well as antibacterial activity have been confirmed among others [11,12].

The phenolic compounds in maqui can be divided into three groups: phenolic acids (gallic acid, hexahydroxydiphenic acid, granatin B, punicacortin C, etc.), flavonols (myricetin, quercetin, kaempferol, and its derivatives), and eight anthocyanins (delphinidin 3-*O*-sambubioside-5-*O*-glucoside, delphinidin 3,5-*O*-diglucoside, cyanidin 3-*O*-sambubioside-5-*O*-glucoside, cyanidin 3,5-*O*-diglucoside, delphinidin 3-*O*-sambubioside, delphinidin 3-*O*-glucoside, cyanidin 3-*O*-glucoside, and cyanidin 3-*O*-sambubioside) [9,13]. This extraordinary content in bioactive compounds has been granted its recognition as a "superfruit" [14].

However, despite the fact that maqui's antioxidant capacity is much higher than that of other fruits, large scale plantations are yet to be found, since it is mainly consumed worldwide as an extract or supplement instead of as fresh fruit.

Moreover, maqui berries have only started to be commercialized and they are understudied to date, so hardly any extraction or analysis techniques have been specifically developed in the literature for this fruit. Due to the different characteristics of this fruit as compared to other similar berries, not only in the matrix, but also in the anthocyanins and phenolic compounds, the development, and optimization of extraction techniques specifically for maqui are required. In addition, due to its high cost, the use of other cheaper berries to replace maqui seems to be a serious problem for the food industry, developing adequate techniques that allow for us to control its quality and detecting possible food fraud is essential [15,16]. Due to their advantages when compared to conventional methods, more environmentally friendly, faster, cheaper, and more energy efficient methods, such as microwave-assisted extraction, ultrasonic-assisted extraction, supercritical fluid extraction, or pressurized liquid extraction, have been used to obtain extracts from plant materials [17,18].

In fact, ultrasound assisted extraction (UAE) has been used for the extraction of compounds of interest, since it is simple, rapid, low cost, and energy efficient [19]. Ultrasounds are very high frequency pressure waves, which are transmitted by materials, causing their contraction and subsequent expansion, and consequently the transmission of energy through those materials. The ultrasound signals generate physical and chemical changes in the medium, as they generate bubbles that subsequently collapse due to cavitation. This cavitation breaks the plant matrix cell walls and it favours the penetration of the solvent and the release of the analytes [20,21]. The solid and liquid particles vibrate and then accelerate, because of the ultrasonic action and, as a result, the solute rapidly passes from the solid phase to the solvent [22]. One of the main advantages of this method is its efficiency, since it can obtain greater extraction yields with a lower solvent consumption and in a shorter time that any other extraction technique [23]. This is a widely used technique that has been recently applied to the extraction of compounds of interest from other similar matrices, such as grapes [24], mulberries [25], or blueberries [26].

Several variables can affect efficiency levels of ultrasonic extraction, such as cycle, amplitude, temperature, or type and volume of solvent as well as its pH. In relation to the solvent, aqueous alcohol mixtures are the most commonly used solvents for the extraction of bioactive components from berries [27]. Temperature and pH can cause the degradation of these compounds, so they are to be kept under control [28,29]. Regarding the cycle, the amplitude or power may favour the destruction of the cell walls and improve the mass transfer during the extraction process [30]. Therefore, a study that is based on a Box-Behnken design was carried out to determine the optimal conditions for the extraction method and to evaluate the importance of each factor and the relationships between them [31,32]. The results were treated with a response surface method; a technique that generates a mathematical model where the response of the system can be observed in terms of the factors that are involved and the interactions between them [23,33].

The aim of this work is to develop, based on the comparison of multiple extraction variables, a green and efficient method to extract compounds of biological interest from maqui berries. The medicinal uses of this berry that are attributed by the plant's secondary metabolites, the unique resources for pharmaceuticals, food additives, and fine chemicals, as well as the high commercial value and demand make the development of simple extraction techniques from this fruit of great interest for the food industry. Moreover, a comparison between UAE and microwave assisted extraction (MAE) performance was also carried out to determine the impact of cavitation in UAE and microwaves in closed systems in MAE.

2. Materials and Methods

2.1. Sample Preparation

Lyophilized maqui from organic farming that was purchased from SuperAlimentos, Mundo Arcoíris, (Besalú, Girona, Spain) was the biological material used for the experiments. The samples were in powder form to increase their contact surface with the solvent and improve the yields. Once the extraction methods had been optimized, several currently commercialized samples in different formats, including capsules, pills, and lyophilized matrix, which contained maqui, were also tested to verify the suitability of the method. Both, the experimental maqui sample and the commercial samples were stored at −20 °C prior to their analysis.

2.2. Chemicals and Reagents

The solvents that were used for the extractions were mixtures of methanol and water. The methanol employed (Fischer Scientific, Loughborough, UK) was HPLC grade. A Milli-Q water purification system from Millipore supplied ultra-pure water (Bedford, MA, USA). To adjust the pH, solutions of hydrochloric acid, and sodium hydroxide (Panreac Química S.A.U., Castellar del Vallés, Barcelona, Spain), grade "for analysis" were used. For the chromatographic separations of the anthocyanins, methanol, milli-Q water, and formic acid (Scharlau S.L., Sentmenat, Barcelona, Spain), HPLC grade, were used, and for its quantification, cyanidin chloride was used (Sigma-Aldrich Chemical Co., St Louis, MO, USA) as a standard pattern. For the quantification of the total phenolic compounds, distilled water, Folin-Ciocalteau reagent (Merck KGaA, EMD Millipore Corporation, Darmstadt, Germany), anhydrous sodium carbonate (Panreac Química S.A.U., Castellar del Vallés, Barcelona, Spain), and gallic acid as a standard (Sigma-Aldrich Chemical Co., St Louis, MO, USA) were used.

2.3. Ultrasound-Assisted Extraction Procedure

The extraction was performed by ultrasonic irradiation while using a Probe UP 200 S (Ultraschallprozessor Dr. Hielscher, Gmbh, Berlin, Germany), which allows for the control and modification of the cycle and the amplitude. This system was coupled with a thermostatic bath (FRIGITERM-10, Selecta, Barcelona, Spain) under controlled temperature. The variables that were to be studied in the different experiments were: percentage of methanol (25-50-75%), pH (2-4.5-7), temperature (10-40-70 °C), sample mass/solvent volume (ratio) (10-15-20 mL), cycle (0.2-0.45-0.7 s), and amplitude (30-50-70%).

Approximately, 0.5 grams of lyophilized sample was weighed in a 50 mL "Falcon" and the appropriate type and volume of solvent was added, depending on the experimental design. The "Falcon" was placed inside a double-walled vessel to control the extraction temperature. The extraction was carried out under controlled UAE conditions for 10 min. After the extraction, the extract was centrifuged twice for 5 min at $11,544 \times g$. The supernatant was transferred in both cases to a 25 mL volumetric flask and then made up to the mark with the same solvent. Finally, the extracts were stored in a freezer at -20 °C prior to their analysis.

2.4. Identification of Anthocyanins

First, the anthocyanins in the extract were filtered through a 0.22-μm syringe filter (Nylon Syringe Filter, FILTER-LAB, Barcelona, Spain) and they were then identified by means of ultra-high-performance liquid chromatography equipment (UHPLC) coupled to a quadrupole-time-of-flight mass spectrometer (QToF-MS) (Xevo G2 QToF, Waters Corp., Milford, MA, USA). The separation was carried out using a C-18 analytical column (Acquity UHPLC BEH C18, Waters Corporation, Milford, MA, USA) of 100 mm × 2.1 mm and 1.7 μm particle size, working in reverse phase. The flow rate was 0.4 mL/min. The mobile phase was a binary solvent system consisting on Milli-Q water that was acidified with 2% formic acid as phase A and pure methanol as phase B, both filtered and degassed, and using the following gradient: 0 min, 15% B; 3.30 min, 20% B; 3.86 min, 30% B; 5.05 min, 40% B; 5.35 min, 55% B; 5.64 min, 60% B; 5.94 min, 95% B; 7.50 min, 95% B. For the determination of the analytes, an electrospray source operating in the positive ionization mode was used under the following conditions: desolvation gas flow = 700 L/h, desolvation temperature = 500 °C, cone gas flow = 10 L h^{-1}, source temperature = 150 °C, capillary voltage = 700 V, cone voltage = 30 V, and collision energy = 20 eV. Full-scan mode was used (m/z = 100–1200). Under the above conditions, eight anthocyanins were identified in the maqui samples (compound, m/z): delphinidin 3-*O*-sambubioside-5-*O*-glucoside, 759; delphinidin 3,5-*O*-diglucoside, 627; cyanidin 3-*O*-sambubioside-5-*O*-glucoside, 743; cyanidin 3,5-*O*-diglucoside, 611; delphinidin 3-*O*-sambubioside, 597; delphinidin 3-*O*-glucoside, 465; cyanidin 3-*O*-glucoside, 449; and, cyanidin 3-*O*-sambubioside, 581.

2.5. Detection of Anthocyanins

Once the anthocyanins were identified in the maqui samples, their separation and quantification was carried out by a liquid chromatography equipment Elite LaChrom Ultra System (VWR Hitachi, Tokyo, Japan), which was composed by an L-2200 U autosampler, an L-2300 column oven set at 50 °C, two L-2160 U pumps, and a UV-Vis L-2420 U detector set at 520 nm, which is the maximum absorption of anthocyanins. The UHPLC chromatogram and the information about each peak assignment are shown in Figure S1 and Table S1, respectively.

As aforementioned, the extracts that were obtained were first filtered through a 0.22 μm syringe filter (Nylon Syringe Filter, FILTER-LAB, Barcelona, Spain). A C-18 column (Phenomenex Kinetex, CoreShell Technology, Torrance, CA, USA) of 100 × 2.1 mm and particle size of 2.6 μm was used, working in reverse phase. The injection volume was 15 μL. For its separation, Milli-Q water acidified at 5% with formic acid (solvent A) and pure methanol (solvent B) were used, both being filtered through a 0.22 μm filter and degassed by ultrasonic bath (Elma S300 Elmasonic, Singen, Germany). The gradient used was as follows: 0.0 min, 2% B; 2.0 min, 2% B; 3.5 min, 15% B; 5.5 min, 25% B; 6.5 min, 40% B; 7.0 min 100% B; 9.3 min, 100% B; 10.0 min, 2% B; 12.0 min, 2% B, and a flow of 0.7 mL/min. For its quantification, a calibration curve was generated, using cyanidin chloride as the reference standard between 0.05 and 30 mg L^{-1}. The regression equation ($y = 252,638.09x - 28,465.10$) and the correlation coefficient (0.9998), as well as the detection and quantification limits (LOD = 0.179 mg L^{-1} and LOQ = 0.597 mg L^{-1}, respectively), were calculated. Finally, assuming that the different anthocyanins have similar absorbance, the anthocyanins that were present in maqui were quantified from the calibration curve of cyanidin chloride, based on the structural similarities and taking their corresponding molecular weights into account. A calibration curve was generated for each anthocyanin present in the maqui berry, which allows for the quantification of each of them [34–36]. The results were expressed as a sum of individual anthocyanins as milligrams of cyanidin chloride equivalents per gram of dry fruit.

2.6. Total Phenolic Content (TPC)

The total phenolic content in maqui was expressed as mg of gallic acid equivalents per gram of fresh fruit, according to the modified Folin-Ciocalteau (FC) spectrophotometric method [37–40]. For this, a UV-Vis Helios Gamma (γ) Unicam (Thermo Fisher Scientific, Waltham, MA, USA) spectrophotometer was used. Prior to their analysis, the extracts were filtered through a 0.45 μm syringe filter (Nylon Syringe Filter, FILTER-LAB, Barcelona, Spain). Subsequently, 0.25 mL of extract, 12.5 mL of water, 1.25 mL of Folin-Ciocalteau reagent, and 5 mL of 20% anhydrous sodium carbonate solution were added to a 25 mL volumetric flask and the solution was made up to the mark with water. A blue colour complex was formed as the result of the reduction of the phenolic compounds that were present in the extract, and after 30 min, its absorbance was determined at 765 nm. The total phenolic content was calculated by means of a calibration curve under the same conditions, using standards of gallic acid of known concentration between 100 and 2000 mg L^{-1} and measuring their absorbance values. The regression equation ($y = 0.0010x + 0.0065$) and the correlation coefficient ($R^2 = 0.9998$) were obtained. The results were expressed as mg of gallic acid equivalent per gram of dry fruit.

2.7. Response Surface Regression Analysis

A three-level, six factors Box-Behnken design (BBD), in conjunction with the surface response method, was employed to determine the optimum UAE conditions for the extractions of both types of compounds and their interactions. BBD is a spherical design structure that prevents carrying out the experiments under extreme conditions [41]. BBD ensures the maximum possible amount of data on the system's response, while a lower number of experiments and smaller amounts of reagent are required [42]. The extraction variables were the percentage of methanol, pH, temperature, solvent volume: sample mass (ratio), cycle, and amplitude, which were coded at three different levels: −1 (low), 0 (medium), and +1 (high). Therefore, the design indicates the execution of 54 experiments, which were randomly carried out to avoid any preconceptions. The experimental data from the two responses—total anthocyanins and total phenolic compounds—were fitted into a second-order polynomial model, as in the following equation [43]:

$$y = \beta_0 + \sum_{i=1}^{k} \beta_i \cdot x_i + \beta_{ii} \cdot x_i^2 + \sum_{i} \sum_{i=1}^{k} \beta_{ij} \cdot x_i x_j + r \tag{1}$$

where, *y* represents the aforementioned responses; β_0 is the constant coefficient; β_i, β_{ii}, and β_{ij} are the regression coefficients of linear, quadratic, and interactive terms respectively; x_i represent each factor; and, *r* is the residual value.

2.8. Statistical Analysis

The results were analyzed by means of the statistical program Design Expert software (Trial Version, Stat-Ease Inc., Minneapolis, MN, USA). An Analysis of Variance (ANOVA) was performed to evaluate the quality of the model fitted to the experimental response, the regression terms, and to determine any statistically significant differences ($p < 0.05$). A response surface method was employed to determine the optimum extraction conditions and the most influential parameters.

3. Results and Discussion

3.1. Fitting the Model of the Extraction Process

For the three-level Box-Behnken design that was employed to determine the optimal UAE conditions, six independent variables: percentage of methanol (25-50-75%), temperature (10-40-70 °C), amplitude (30-50-70%), cycle (0.2-0.45-0.7 s), pH (2-4.5-7), and solvent volume:sample mass ratio (10:0.5-15:0.5-20:0.5 mL:g) and two responses: total anthocyanins and total phenolic compounds, were optimized. The decoded values of the independent variables and the responses that were obtained in the multivariate study from each experiment are shown in Table 1.

Table 1. Box-Behnken design matrix including both decoded variables and responses.

	Factors						Responses	
Run	Solvent X_1	Temp.* X_2	Amplitude X_3	Cycle X_4	pH X_5	Ratio X_6	Total Anthocyanins (mg g^{-1})	Total Phenolic Compounds (mg g^{-1})
1	50	40	30	0.45	2	10	34.21	37.10
2	50	40	70	0.45	2	10	34.48	38.49
3	50	40	30	0.45	7	10	31.29	44.45
4	50	40	70	0.45	7	10	31.00	47.68
5	50	40	30	0.45	2	20	38.17	50.36
6	50	40	70	0.45	2	20	40.00	55.15
7	50	40	30	0.45	7	20	33.19	48.39
8	50	40	70	0.45	7	20	31.88	51.44
9	50	10	50	0.2	2	15	29.96	38.07
10	50	70	50	0.2	2	15	36.47	60.46
11	50	10	50	0.7	2	15	33.86	35.18
12	50	70	50	0.7	2	15	38.77	71.07
13	50	10	50	0.2	7	15	30.73	38.89
14	50	70	50	0.2	7	15	35.54	46.32
15	50	10	50	0.7	7	15	31.35	41.21
16	50	70	50	0.7	7	15	31.12	44.11
17	25	40	30	0.2	4.5	15	21.11	39.51
18	75	40	30	0.2	4.5	15	31.95	37.39
19	25	40	70	0.2	4.5	15	21.20	40.57
20	75	40	70	0.2	4.5	15	31.22	38.26
21	25	40	30	0.7	4.5	15	21.20	41.96
22	75	40	30	0.7	4.5	15	33.66	46.65
23	25	40	70	0.7	4.5	15	25.17	45.97
24	75	40	70	0.7	4.5	15	35.10	39.14
25	50	10	30	0.45	4.5	10	30.53	47.25
26	50	70	30	0.45	4.5	10	34.19	37.92
27	50	10	70	0.45	4.5	10	26.92	46.70
28	50	70	70	0.45	4.5	10	32.26	45.80
29	50	10	30	0.45	4.5	20	31.11	36.72
30	50	70	30	0.45	4.5	20	36.91	42.25
31	50	10	70	0.45	4.5	20	27.17	41.08
32	50	70	70	0.45	4.5	20	37.19	41.07
33	25	10	50	0.45	2	15	21.56	35.93
34	75	10	50	0.45	2	15	35.88	44.42
35	25	70	50	0.45	2	15	28.78	51.74
36	75	70	50	0.45	2	15	36.10	56.54
37	25	10	50	0.45	7	15	21.79	35.11
38	75	10	50	0.45	7	15	35.34	37.37
39	25	70	50	0.45	7	15	24.80	47.58
40	75	70	50	0.45	7	15	33.68	46.77
41	25	40	50	0.2	4.5	10	20.26	37.54
42	75	40	50	0.2	4.5	10	31.89	39.78
43	25	40	50	0.7	4.5	10	19.91	38.76
44	75	40	50	0.7	4.5	10	32.76	48.22
45	25	40	50	0.2	4.5	20	20.08	37.56
46	75	40	50	0.2	4.5	20	34.21	39.58
47	25	40	50	0.7	4.5	20	23.60	43.52
48	75	40	50	0.7	4.5	20	34.09	47.47
49	50	40	50	0.45	4.5	15	33.66	43.97

Table 1. Cont.

Run	Factors						Responses	
	Solvent X_1	Temp.* X_2	Amplitude X_3	Cycle X_4	pH X_5	Ratio X_6	Total Anthocyanins (mg g^{-1})	Total Phenolic Compounds (mg g^{-1})
50	50	40	50	0.45	4.5	15	31.30	46.44
51	50	40	50	0.45	4.5	15	34.37	44.78
52	50	40	50	0.45	4.5	15	33.64	45.63
53	50	40	50	0.45	4.5	15	29.16	44.08
54	50	40	50	0.45	4.5	15	31.90	46.76

* Temp.: Temperature; X_1: Percentage of methanol; X_2: Temperature; X_3: Amplitude; X_4: Cycle; X_5: pH; X_6: ratio "Sample mass/volume of solvent".

Analysis of variance (ANOVA) validates the suitability of the model. This allows for evaluating the effect of the variables to identify the possible interactions between them and to assess the statistical significance of the model, whose results can be seen in Table 2. This analysis also provides information on the mathematical model that is generated from the experimental data. Once the 54 experiments were carried out, the coefficients for the full second-order polynomial equation for both types of compounds were established to predict the responses. In this way, two suitable mathematical models were obtained to describe the response values of the anthocyanins (Y_{TA}) and phenolic compounds (Y_{TP}), as a function of the independent variables.

$$Y_{TA} \text{ (mg g}^{-1}) = 32.34 + 5.68X_1 + 2.07X_2 - 0.16X_3 + 0.66X_4 - 1.52X_5 + 1.16X_6 - 4.35X_1{}^2 - 1.46X_{1 \times 2} - 0.42X_{1 \times 3} - 0.06X_{1 \times 4} + 0.10X_{1 \times 5} + 0.02X_{1 \times 6} - 0.24X_2{}^2 + 0.74X_{2 \times 3} - 0.83X_{2 \times 4} - 0.81X_{2 \times 5} + 0.85X_{2 \times 6} + 0.21X_3{}^2 + 0.76X_{3 \times 4} - 0.46X_{3 \times 5} + 0.15X_{3 \times 6} - 0.62X_4{}^2 - 1.25X_{4 \times 5} + 0.36X_{4 \times 6} + 2.00X_5{}^2 - 0.84X_{5 \times 6} - 0.27X_6{}^2 \quad (2)$$

$$Y_{TP} \text{ (mg g}^{-1}) = 45.28 + 1.08X_1 + 4.74X_2 + 0.89X_3 + 2.05X_4 - 1.88X_5 + 1.04X_6 - 2.76X_1{}^2 - 0.84X_{1 \times 2} - 1.46X_{1 \times 3} + 0.71X_{1 \times 4} - 1.48X_{1 \times 5} - 0.72X_{1 \times 6} - 1.18X_2{}^2 + 0.36X_{2 \times 3} + 1.12X_{2 \times 4} - 3.37X_{2 \times 5} + 1.97X_{2 \times 6} - 1.05X_3{}^2 - 0.68X_{3 \times 4} + 0.01X_{3 \times 5} - 0.06X_{3 \times 6} - 0.28X_4{}^2 - 0.95X_{4 \times 5} + 0.52X_{4 \times 6} + 3.01X_5{}^2 - 2.78X_{5 \times 6} - 0.69X_6{}^2 \quad (3)$$

Table 2. Analysis of variance (ANOVA) of the quadratic model adjusted to the extraction yield. (**A**) Total anthocyanins; and, (**B**) Total phenolic compounds.

(A)						
Source	Degrees of Freedom	Sum of Squares	Mean Square	F-Value	P-Value	Coefficient
Model	27	1429.49	52.94	18.40	0.0000	
Intercept	1					32.34
X_1	1	775.46	775.46	269.44	0.0000	5.68
X_2	1	102.53	102.53	35.62	0.0000	2.07
X_3	1	0.6324	0.6324	0.2197	0.6431	−0.1623
X_4	1	10.63	10.63	3.69	0.0657	0.6654
X_5	1	55.61	55.61	19.32	0.0002	−1.52
X_6	1	32.51	32.51	11.29	0.0024	1.16
$X_{1 \times 2}$	1	17.06	17.06	5.93	0.0221	−1.46
$X_{1 \times 3}$	1	1.40	1.40	0.4851	0.4923	−0.4178
$X_{1 \times 4}$	1	0.0503	0.0503	0.0175	0.8958	−0.0561
$X_{1 \times 5}$	1	0.0783	0.0783	0.0272	0.8703	0.0989
$X_{1 \times 6}$	1	0.0027	0.0027	0.0009	0.9759	0.0183
$X_{2 \times 3}$	1	4.34	4.34	1.51	0.2304	0.7366
$X_{2 \times 4}$	1	5.52	5.52	1.92	0.1780	−0.8304
$X_{2 \times 5}$	1	10.44	10.44	3.63	0.0680	−0.8077
$X_{2 \times 6}$	1	5.80	5.80	2.02	0.1676	0.8515

Table 2. *Cont.*

(A)

Source	Degrees of Freedom	Sum of Squares	Mean Square	F-Value	P-Value	Coefficient
$X_{3 \times 4}$	1	4.58	4.58	1.59	0.2181	0.7570
$X_{3 \times 5}$	1	1.71	1.71	0.5942	0.4477	−0.4624
$X_{3 \times 6}$	1	0.3622	0.3622	0.1259	0.7256	0.1505
$X_{4 \times 5}$	1	12.55	12.55	4.36	0.0467	−1.25
$X_{4 \times 6}$	1	1.03	1.03	0.3591	0.5542	0.3594
$X_{5 \times 6}$	1	5.60	5.60	1.95	0.1749	−0.8366
X_1^2	1	194.81	194.81	67.69	0.0000	−4.35
X_2^2	1	0.6133	0.6133	0.2131	0.6482	−0.2442
X_3^2	1	0.4469	0.4469	0.1553	0.6968	0.2084
X_4^2	1	3.92	3.92	1.36	0.2539	−0.6172
X_5^2	1	41.13	41.13	14.29	0.0008	2.00
X_6^2	1	0.7366	0.7366	0.2559	0.6172	−0.2676
Residual	26	74.83	2.88			
Lack of fit	21	55.92	2.66	0.7042	0.7421	
Pure error	5	18.91	3.78			
Total	53	1504.32				

(B)

Source	Degrees of Freedom	Sum of Squares	Mean Square	F−Value	P−Value	Coefficient
Model	27	1417.49	52.50	1.41	0.1926	
Intercept	1					45.28
X_1	1	27.83	27.83	0.7467	0.3954	1.08
X_2	1	538.78	538.78	14.46	0.0008	4.74
X_3	1	19.06	19.06	0.5114	0.4809	0.8911
X_4	1	101.19	101.19	2.72	0.1114	2.05
X_5	1	85.09	85.09	2.28	0.1428	−1.88
X_6	1	25.79	25.79	0.6920	0.4131	1.04
$X_{1 \times 2}$	1	5.72	5.72	0.1535	0.6984	−0.8455
$X_{1 \times 3}$	1	17.15	17.15	0.4604	0.5035	−1.46
$X_{1 \times 4}$	1	8.19	8.19	0.2199	0.6431	0.7156
$X_{1 \times 5}$	1	17.47	17.47	0.4689	0.4995	−1.48
$X_{1 \times 6}$	1	4.11	4.11	0.1103	0.7425	−0.7168
$X_{2 \times 3}$	1	1.04	1.04	0.0278	0.8688	0.3600
$X_{2 \times 4}$	1	10.06	10.06	0.2699	0.6078	1.12
$X_{2 \times 5}$	1	182.28	182.28	4.89	0.0360	−3.38
$X_{2 \times 6}$	1	31.00	31.00	0.8318	0.3701	1.97
$X_{3 \times 4}$	1	3.69	3.69	0.0990	0.7555	−0.6791
$X_{3 \times 5}$	1	0.0017	0.0017	0.0000	0.9947	0.0145
$X_{3 \times 6}$	1	0.0520	0.0520	0.0014	0.9705	−0.0570
$X_{4 \times 5}$	1	7.23	7.23	0.1940	0.6632	−0.9506
$X_{4 \times 6}$	1	2.21	2.21	0.0592	0.8096	0.5252
$X_{5 \times 6}$	1	61.66	61.66	1.65	0.2097	−2.78
X_1^2	1	78.29	78.29	2.10	0.1592	−2.76
X_2^2	1	14.43	14.43	0.3873	0.5391	−1.18
X_3^2	1	11.50	11.50	0.3086	0.5833	−1.06
X_4^2	1	0.8004	0.8004	0.0215	0.8846	−0.2790
X_5^2	1	98.81	98.81	2.65	0.1155	3.10
X_6^2	1	4.86	4.86	0.1305	0.7208	−0.6876
Residual	26	968.87	37.26			
Lack of fit	21	961.79	45.80	32.34	0.0006	
Pure error	5	7.08	1.42			
Total	53	2386.36				

As far as the total anthocyanins are concerned, the factors that had a significant linear influence on the response, with a 95% level of confidence, were the percentage of methanol, temperature, pH, and ratio, since their *p*-values were lower than 0.05. Besides, the lack of fit test showed a *p*-value that is higher than 0.05 (not significant), which means that the model fits well.

With regards to the phenolic compounds, the influential variables with a *p*-value lower than 0.05 were temperature and the interaction temperature-pH. In this case, the lack of fit test was significant, with a *p*-value < 0.05, which indicates that the regression seemed to be inadequate. However, it must be taken into account that phenolic compounds are a group of molecules of high diversity, with a wide range in terms of polarity and sizes. Therefore, the optimal conditions that were determined are a compromise status, where the greatest amount of the desired compounds can be extracted [44].

The Pareto charts in Figure 1 represent the significant effects of all the variables, both linear and quadratic, as well as their interactions. The effects are displayed in decreasing order of significance. The length of each bar is proportional to the absolute magnitude of the estimated effects coefficients, while the vertical line represents the minimal magnitude of statistically significant effects (95% confidence level) with respect to the response.

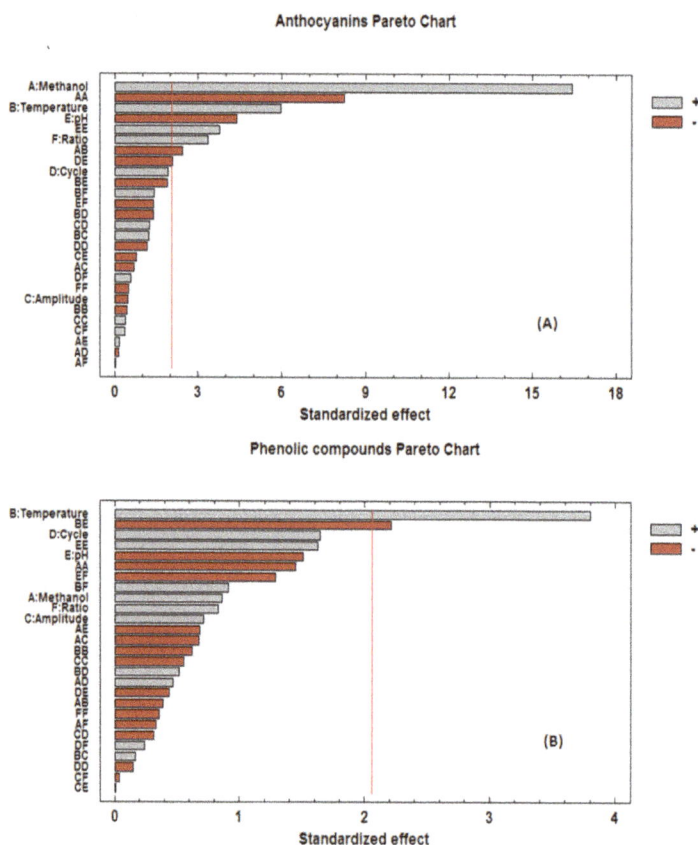

Figure 1. Pareto charts for the standardized effects: (**A**) total anthocyanins and (**B**) total phenolic compounds.

Regarding anthocyanins, the variable that had the greatest effect on the response was the percentage of methanol, which was determined by the polarity characteristics of both the solvent and the compounds that are present in maqui. Temperature, pH, and ratio also had significant effects, although to a less extent. The percentage of methanol, temperature, and ratio had a positive effect, which means that an increase in these factors favoured the recovery of anthocyanins in the extract. Higher extraction temperatures accelerate molecular motion, penetration, dissolution, and diffusion to favour the releasing of anthocyanins [45]. Regarding the ratio, a smaller amount of sample in a large volume of solvent (large variation in concentration) leads to a greater gradient and, therefore, to a greater extraction, favoured by the transfer of mass. On the contrary, pH has a negative effect on the response variable, i.e., the extraction of anthocyanins was more successful when the pH values were low. The recovery of the anthocyanins in an acidic medium increases due to their stable conformation thanks to the cation flavylium, which confers them their red colour [46,47]. Moreover, the percentage of methanol and pH had a significant quadratic influence on the response, as well as on the interactions methanol-temperature and pH-cycle. Once the influence of the different factors was known, the reduced second order polynomial equation was obtained, where only those variables and/or interactions that had shown a significant effect on the response were considered:

$$Y_{TA} = 32.34 + 5.68X_1 + 2.07X_2 - 1.52X_5 + 1.16X_6 - 4.35X_1{}^2 - 1.46X_{1 \times 2} - 1.25X_{4 \times 5} + 2.00X_5{}^2 \quad (4)$$

As far as the total phenolic compounds are concerned, the Pareto chart that was obtained differs slightly from the previous one, since fewer influential variables were observed. The analysis of the model clearly showed that temperature was the most influential factor, with a marked positive effect on the extraction. High temperatures favour the breakage of Van der Waals, hydrogen, molecular, or dipole-dipole bonds between the compounds to be extracted, which in turn reduced the required energy that is necessary for their desorption. In addition, the viscosity and the surface tension of the solvent decreases at higher temperatures, which improves the penetration of the solvent into the matrix and a faster dissolution of the extract. This results in the subsequent increase in mass transfer and a greater overall yield [48,49]. The rest of the variables had a similar effect, but less significant effect in absolute terms, so they were not significant factors. Again, a simplified second order polynomial equation was obtained and the results were similar to those of the complete equation:

$$Y_{TP} = 30.19 + 3.16X_2 - 2.25X_{2 \times 5} \quad (5)$$

3.2. Optimal Conditions

After the statistical treatment of the data, the optimization of the variables was evaluated while using the quadratic mathematical model within the experimental range studied. Table 3 shows the optimal UAE conditions for the extraction of both anthocyanins and phenolic compounds.

Table 3. Optimal conditions for ultrasound-assisted extraction. (A) Total anthocyanins; and, (B) Total phenolic compounds.

	(A) Total Anthocyanins	(B) Total Phenolic Compounds
Percentage of methanol (%)	61.5	50
pH	2.1	2
Ratio (mL:g)	20:0.5	20:0.5
Temperature (°C)	69.4	70
Amplitude (%)	46	35
Cycle (s)	0.7	0.7

Firstly, it can be observed that, in both cases, the optimum percentage of methanol was an intermediate value between 45–65%. Several papers report an increase in anthocyanins and phenolic compounds extraction with sonication in moderately polar media, which can be explained by the degradation of cellular walls as a result of the cavitation bubbles collapse. Therefore, these compounds are removed from their original location and then transferred to the solvent volume, which favours greater extraction yields [46,50,51].

Temperature is an influential factor in the extraction of bioactive compounds that may affect the stability of phenolic compounds [52]. Using higher temperatures was discarded, because it may reduce the recovery, since these compounds are thermally sensitive, and thus can be easily degraded by the hydrolysis of glucoside compounds and form their corresponding unstable aglycones or by the hydrolytic opening of the heterocyclic ring that would form chalcone in the case of anthocyanins [53]. In addition, working at over 70 °C was not recommended, since methanol may undergo a change of phase and it can evaporate.

Regarding pH, levels that were below 2 were not considered to work, since the compounds of interest might undergo acid hydrolysis [29].

Regarding the ratio, the maximum volume of the range studied was the optimal value. No higher ratios were considered, since the compounds would be difficult to quantify when they are below their quantification limits.

Finally, although the energy that was provided by ultrasound is necessary to release the compounds from the matrix, it may also accelerate the degradation process of the phenolic compounds [54]. In general, the action of the ultrasounds would increase the solubility of the molecules by destroying the intramolecular and intermolecular bonds and by increasing the contact between the hydrophilic groups and the extraction solvents [55]. Although it is clear that the cycle also presents an extreme value, its modification was not considered, since it was not an influential variable.

When the above-mentioned optimal conditions were applied, the total average concentrations of anthocyanins and phenolic compounds were 33.02 mg g^{-1} and 50.26 mg g^{-1}, respectively. Similar results were reported by other authors in some vegetable matrices [8,56]. It can be observed that the main phenolic compounds that are present in maqui berries are anthocyanins, unlike many other berries that have less anthocyanin content in comparison with the total phenolic compounds. This high anthocyanin's content greatly contributes to their superior antioxidant capacity [7].

3.3. Extraction Kinetics

Time is another significant variable in the extraction of anthocyanins and phenolic compounds from different matrices [26]. In order to determine its impact on the process, a study of kinetics was carried out under the previously determined optimal UAE conditions. Different extraction times of 2, 5, 10, 15, 20, and 25 min were applied to the extraction process and each test was carried out in triplicate. The resulting recovery of anthocyanins and the total phenolic compounds are shown in Figure 2. In the case of the anthocyanins, it was observed that after 5 min there were no statistically significant differences. Thus, the extraction time was determined at 5 min, since it means saving both time and money. A different trend was obtained for total phenolic compounds, where the recovery reached its maximum at 15 min. In that case, a longer time was needed for the extraction of phenolic compounds.

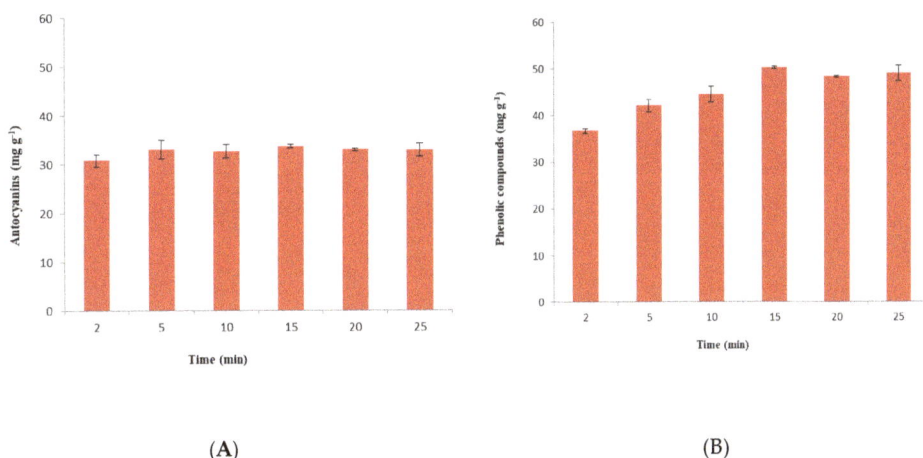

(A)

(B)

Figure 2. Effect of the extraction time on the recovery. (**A**) Total anthocyanins; and, (**B**) Total phenolic compounds.

3.4. Repeatability and Intermediate Precision

The reliability of the developed UAE methods was evaluated by repeatability and intermediate precision. The former was evaluated by 12 extractions on the same day and the latter based on 12 extractions per day during three consecutive days. A total of 36 extractions were performed under the optimal conditions that were previously determined. The results were expressed by the coefficient of variation, which was 2.48% and 3.37% in the case of repeatability, and 2.92% and 3.95% in the case of intermediate precision, for anthocyanins and phenolic compounds, respectively. These results, with coefficients lower than 5%, are within the acceptable limits that were defined by the Association of Official Analytical Chemists (AOAC) [57] and indicate satisfactory repeatability and intermediate precision of the method.

3.5. Application to Real Samples

The suitability of the newly developed methods was evaluated by applying them to the extraction of seven real samples (capsules (M-1 and M-2), pills (M-3), and lyophilized maqui (M-4, M-5, M-6, and M-7)) under the previously determined optimum conditions. The analyses were carried out in triplicate and Table 4 shows the extracted compounds. For the quantification of total anthocyanins and total phenolic compounds, the extracts were analyzed by UHPLC and the spectrophotometric method of Folin-Ciocalteau, respectively. The results from the different samples show the same trend for the extraction of anthocyanins and total phenolic compounds. Despite being M-1 and M-2 capsules of maqui, the highest concentration was obtained from M-2 and the lowest one for M-1. This may be due to the concentration of the raw material used. It can also be due to the different thermal or storage treatment of the samples. Both can cause changes in physical and nutritional properties, as well as the significant degradation of these compounds, particularly anthocyanins [58,59]. On the other hand, it should be noted that no anthocyanins were extracted from M-3. This is because only two anthocyanins (cyanidin 3-*O*-glucoside and delphinidin 3-*O*-glucoside) were detected, instead of the eight characteristic anthocyanins that were found in maqui. Therefore, M-3's composition does not match that of the berries analyzed in this work. Based on these results, how important it is to count on the right extraction techniques and the adequate equipment to prevent any food fraud, and to ensure the quality of the product to consumers has been confirmed. Finally, the amount of compounds of biological interest extracted from the different lyophilized samples (M 4-7) was very similar and relatively high, since the lyophilized samples preserve their beneficial properties and characteristics during their storage and transport [60].

Table 4. Total anthocyanins (mg g^{-1}) and total phenolic compounds (mg g^{-1}) extracted for each commercial samples by means of ultrasound assisted extraction (UAE) and microwave assisted extraction (MAE).

Foodstuff Made with Maqui	Total Anthocyanins (mg g^{-1})		Total Phenolic Compounds (mg g^{-1})	
	UAE	MAE	UAE	MAE
M-1	2.13 ± 0.16 [a]	1.73 ± 0.16 [b]	6.83 ± 0.23 [a]	8.22 ± 0.34 [b]
M-2	78.73 ± 0.67 [a]	75.55 ± 3.80 [a]	100.27 ± 1.44 [a]	103.30 ± 0.30 [a]
M-3	-*	-*	10.46 ± 0.29 [a]	11.45 ± 0.45 [b]
M-4	30.81 ± 3.13 [a]	30.35 ± 3.25 [a]	47.25 ± 1.44 [a]	53.06 ± 1.53 [b]
M-5	28.39 ± 1.72 [a]	27.66 ± 1.02 [a]	43.40 ± 1.99 [a]	49.29 ± 2.17 [b]
M-6	37.26 ± 2.08 [a]	35.51 ± 1.40 [a]	58.28 ± 1.33 [a]	59.57 ± 0.70 [a]
M-7	23.21 ± 0.73 [a]	19.89 ± 1.44 [b]	50.95 ± 1.38 [a]	52.13 ± 1.44 [a]

* The characteristic eight anthocyanins found in maqui were not detected. The use of different letters on the same line for each kind of compound indicates that the means differs significantly according to Tukey's test ($p < 0.05$).

The results obtained by means of UAE were compared to those achieved by MAE methods. The samples were analyzed according to the optimal conditions determined in a previous work [61]. Approximately half of the commercial samples analyzed show statistically significant differences with regards to total phenolic compounds content. In turn, an increase in the extraction of phenolic compounds was observed when MAE was employed. This improved performance could be attributed to the ability of microwaves to penetrate into the cell matrix and cause strong absorption by polar molecules, which in turn produces an increase in temperature and pressure within the plant cell. Such an increase in pressure leads to the rupture of cell walls and to the release of analytes [62]. In addition, the decomposition of larger phenolic compounds into smaller ones with intact properties may lead to a greater yield, according to the Folin-Ciocalteau assay [63]. It should be noted that, when optimum conditions are applied to UAE for the extraction of phenolic compounds, the solvent demand is reduced (58% methanol for UAE vs. 65% methanol for MAE). In relation to each method's efficiency to extract anthocyanins, only two of the samples presented a significant greater yield when UAE was applied. This may be because anthocyanins are extremely sensitive to degradation when exposed to high pressure and high temperature, and this has a negative impact on their recovery [64,65]. Finally, it is noteworthy to highlight that UAE is considered to be a fast, effective, and economic alternative, as long as industrial scale processing equipment is designed to obtain greater benefits [66]. Some of its relevant advantages are more rapid kinetics and a higher extraction efficiency and performance [67]. Therefore, the results that were obtained in this study show that UAE can be considered as an efficient alternative and a powerful means for the extraction of both anthocyanins and phenolic compounds.

4. Conclusions

Ultrasound-assisted extraction proved to be quite an effective and fast technique in the extraction of a wide range of polyphenols from maqui berries. BBD was also successfully used to establish the optimum parameters for the extraction of bioactive compounds. The most influential variables on the extraction of anthocyanins and phenolic compounds were methanol percentage and temperature, respectively. It took a longer time for optimal extraction of phenolic compounds, being 15 min, while for anthocyanins, it was only 5 min. Excellent repeatability and intermediate precision were found, with values that were lower than 5%. The developed methods were successfully applied to several commercial samples with maqui content. According to the results, UAE obtained extractions with a greater content in anthocyanins than MAE. It was also observed that, under optimal conditions, UAE used a smaller amount of solvent than MAE. Based on these data, it can be concluded that, when

UAE is applied under optimal conditions, it can be considered as a simple and economical alternative method for the extraction of anthocyanins and total phenolic compounds from maqui.

Supplementary Materials: The following are available online at http://www.mdpi.com/2073-4395/9/3/148/s1, Figure S1: Ultra-high-performance liquid chromatography (UHPLC) chromatogram identified in the ultrasound-assisted extraction, Table S1: HPLC results for each of the anthocyanins found in maqui.

Author Contributions: Conceptualization, E.E.-B. and G.F.B.; methodology, M.V.-E. and A.V.G.d.P.; software, M.F.-G.; formal analysis, C.C.; M.V.-E. and A.V.G.d.P.; investigation, M.V.-E. and A.V.G.d.P.; resources, M.P.; data curation, E.E.-B., M.F.-G. and G.F.B.; writing—original draft preparation, M.V.-E. and A.V.G.d.P.; writing—review and editing, G.F.B. and E.E.-B.; supervision, E.E.-B. and G.F.B.; project administration, G.F.B. and E.E.-B.

Funding: The authors acknowledge V. la Andaluza and University of Cadiz for the support provided through the project OT2016/046.

Conflicts of Interest: The authors declare no conflicts of interest.

References

1. Genskowsky, E.; Puente, L.A.; Pérez-Álvarez, J.A.; Fernández-López, J.; Muñoz, L.A.; Viuda-Martos, M. Determination of polyphenolic profile, antioxidant activity and antibacterial properties of maqui [*Aristotelia chilensis* (Molina) Stuntz] a Chilean blackberry. *J. Sci. Food Agric.* **2016**, *96*, 4235–4242. [CrossRef] [PubMed]

2. Belhachat, D.; Mekimene, L.; Belhachat, M.; Ferradji, A.; Aid, F. Application of response surface methodology to optimize the extraction of essential oil from ripe berries of *Pistacia lentiscus* using ultrasonic pretreatment. *J. Appl. Res. Med. Aromat. Plants* **2018**, *9*, 132–140. [CrossRef]

3. Fredes, C.; Robert, P. The powerful colour of the maqui (*Aristotelia chilensis* [Mol.] Stuntz) fruit. *J. Berry Res.* **2014**, *4*, 175–182. [CrossRef]

4. Gironés-Vilaplana, A.; Mena, P.; García-Viguera, C.; Moreno, D.A. A novel beverage rich in antioxidant phenolics: Maqui berry (*Aristotelia chilensis*) and lemon juice. *LWT Food Sci. Technol.* **2012**, *47*, 279–286. [CrossRef]

5. Fredes, C.; Yousef, G.G.; Robert, P.; Grace, M.H.; Lila, M.A.; Gómez, M.; Gebauer, M.; Montenegro, G. Anthocyanin profiling of wild maqui berries (*Aristotelia chilensis* [Mol.] Stuntz) from different geographical regions in Chile. *J. Sci. Food Agric.* **2014**, *94*, 2639–2648. [CrossRef]

6. Brauch, J.E.; Buchweitz, M.; Schweiggert, R.M.; Carle, R. Detailed analyses of fresh and dried maqui (*Aristotelia chilensis* (Mol.) Stuntz) berries and juice. *Food Chem.* **2016**, *190*, 308–316. [CrossRef] [PubMed]

7. Kumar, S.; Sharma, S.; Vasudeva, N. Review on antioxidants and evaluation procedures. *Chin. J. Integr. Med.* **2017**, *1*, 1–12. [CrossRef]

8. Georgantzi, C.; Lioliou, A.E.; Paterakis, N.; Makris, D.P. Combination of Latic Acid-Based Deep Eutectic Solvents (DES) with β-Cyclodextrin: Performance Screening Using Ultrasound-Assisted Extraction of Polyphenols from Selected Native Greek Medicinal Plants. *Agronomy* **2017**, *7*, 54. [CrossRef]

9. Escribano-Bailón, M.T.; Alcalde-Eon, C.; Muñoz, O.; Rivas-Gonzalo, J.C.; Santos-Buelga, C. Anthocyanins in berries of Maqui (*Aristotelia chilensis* (Mol.) Stuntz). *Phytochem. Anal.* **2006**, *17*, 8–14. [CrossRef]

10. Quispe-Fuentes, I.; Vega-Gálvez, A.; Campos-Requena, V. Antioxidant Compound Extraction from Maqui (*Aristotelia chilensis* [Mol] Stuntz) Berries: Optimization by Response Surface Methodology. *Antioxidants* **2017**, *6*, 10. [CrossRef]

11. Genskowsky, E.; Puente, L.A.; Pérez-Álvarez, J.A.; Fernandez-Lopez, J.; Muñoz, L.A.; Viuda-Martos, M. Assessment of antibacterial and antioxidant properties of chitosan edible films incorporated with maqui berry (*Aristotelia chilensis*). *LWT Food Sci. Technol.* **2015**, *64*, 1057–1062. [CrossRef]

12. Zúñiga, G.E.; Tapia, A.; Arenas, A.; Contreras, R.A.; Zúñiga-Libano, G. Phytochemistry and biological properties of *Aristotelia chilensis* a Chilean blackberry: A review. *Phytochem. Rev.* **2017**, *16*, 1081–1094. [CrossRef]

13. Brauch, J.E.; Reuter, L.; Conrad, J.; Vogel, H.; Schweiggert, R.M.; Carle, R. Characterization of anthocyanins in novel Chilean maqui berry clones by HPLC–DAD–ESI/MSnand NMR-spectroscopy. *J. Food Compos. Anal.* **2017**, *58*, 16–22. [CrossRef]

14. Cespedes, C.L.; Pavon, N.; Dominguez, M.; Alarcon, J.; Balbontin, C.; Kubo, I.; El-Hafidi, M.; Avila, J.G. The chilean superfruit black-berry *Aristotelia chilensis* (*Elaeocarpaceae*), Maqui as mediator in inflammation-associated disorders. *Food Chem. Toxicol.* **2016**, *108*, 438–450. [CrossRef]
15. Manning, L. Food fraud: Policy and food chain. *Curr. Opin. Food Sci.* **2016**, *10*, 16–21. [CrossRef]
16. Van Ruth, S.M.; Huisman, W.; Luning, P.A. Food fraud vulnerability and its key factors. *Trends Food Sci. Technol.* **2017**, *67*, 70–75. [CrossRef]
17. Medina-Torres, N.; Ayora-Talavera, T.; Espinosa-Andrews, H.; Sánchez-Contreras, A.; Pacheco, N. Ultrasound Assisted Extraction for the Recovery of Phenolic Compounds from Vegetable Sources. *Agronomy* **2017**, *7*, 47. [CrossRef]
18. Castejón, N.; Luna, P.; Señoráns, F.J. Alternative oil extraction methods from *Echium plantagineum* L. seeds using advanced techniques and green solvents. *Food Chem.* **2018**, *244*, 75–82. [CrossRef]
19. Bamba, B.S.B.; Shi, J.; Tranchant, C.C.; Xue, S.J.; Forney, C.F.; Lim, L.T. Influence of Extraction Conditions on Ultrasound-Assisted Recovery of Bioactive Phenolics from Blueberry Pomace and Their Antioxidant Activity. *Molecules* **2018**, *23*, 1685. [CrossRef]
20. Espada-Bellido, E.; Ferreiro-González, M.; Carrera, C.; Palma, M.; Barroso, C.G.; Barbero, G.F. Optimization of the ultrasound-assisted extraction of anthocyanins and total phenolic compounds in mulberry (*Morus nigra*) pulp. *Food Chem.* **2017**, *219*, 23–32. [CrossRef]
21. Khan, M.K.; Abert-Vian, M.; Fabiano-Tixier, A.S.; Dangles, O.; Chemat, F. Ultrasound-assisted extraction of polyphenols (flavanone glycosides) from orange (*Citrus sinensis* L.) peel. *Food Chem.* **2010**, *119*, 851–858. [CrossRef]
22. Esclapez, M.D.; García-Pérez, J.V.; Mulet, A.; Cárcel, J.A. Ultrasound-Assisted Extraction of Natural Products. *Food Eng. Rev.* **2011**, *3*, 108–120. [CrossRef]
23. Ali, A.; Lim, X.Y.; Chong, C.H.; Mah, S.H.; Chua, B.L. Optimization of ultrasound-assisted extraction of natural antioxidants from *Piper betle* using response surface methodology. *LWT Food Sci. Technol.* **2018**, *89*, 681–688. [CrossRef]
24. Duan, L.L.; Jiang, R.; Shi, Y.; Duan, C.Q.; Wu, G.F. Optimization of ultrasonic-assisted extraction of higher fatty acids in grape berries (seed-free fruit sections). *Anal. Methods* **2016**, *8*, 6208–6215. [CrossRef]
25. Nguyen, T.N.T.; Phan, L.H.N.; Le, V.V.M. Enzyme-assisted and ultrasound-assisted extraction of phenolics from mulberry (*Morus alba*) fruit: Comparison of kinetic parameters and antioxidant level. *Int. Food Res. J.* **2014**, *21*, 1937–1940.
26. He, B.; Zhang, L.L.; Yue, X.Y.; Liang, J.; Jiang, J.; Gao, X.L.; Yue, P.X. Optimization of Ultrasound-Assisted Extraction of phenolic compounds and anthocyanins from blueberry (*Vaccinium ashei*) wine pomace. *Food Chem.* **2016**, *204*, 70–76. [CrossRef] [PubMed]
27. Rezaei, E.; Abedi, M. Efficient Ultrasound-Assisted Extraction of Cichoric Acid from *Echinacea purpurea* Root. *Pharm. Chem. J.* **2017**, *51*, 471–475. [CrossRef]
28. Patras, A.; Brunton, N.P.; O'Donnell, C.; Tiwari, B.K. Effect of thermal processing on anthocyanin stability in foods; mechanisms and kinetics of degradation. *Trends Food Sci. Technol.* **2010**, *21*, 3–11. [CrossRef]
29. Ross, K.A.; Beta, T.; Arntfield, S.D. A comparative study on the phenolic acids identified and quantified in dry beans using HPLC as affected by different extraction and hydrolysis methods. *Food Chem.* **2009**, *113*, 336–344. [CrossRef]
30. Wen, C.; Zhang, J.; Zhang, H.; Dzah, C.S.; Zandile, M.; Duan, Y.; Ma, H.; Luo, X. Advances in ultrasound assisted extraction of bioactive compounds from cash crops—A review. *Ultrason. Sonochem.* **2018**, *48*, 538–549. [CrossRef]
31. Chen, S.X.; Li, K.K.; Pubu, D.; Jiang, S.P.; Chen, B.; Chen, L.R.; Yang, Z.; Ma, C.; Gong, X.J. Optimization of Ultrasound-Assisted Extraction, HPLC and UHPLC-ESI-Q-TOF-MS/MS Analysis of Main Macamides and Macaenes from Maca (Cultivars of *Lepidium meyenii* Walp). *Molecules* **2017**, *22*, 2196. [CrossRef] [PubMed]
32. Wang, B.; Qu, J.; Luo, S.; Feng, S.; Li, T.; Yuan, M.; Huang, Y.; Liao, J.; Yang, R.; Ding, C. Optimization of Ultrasound-Assisted Extraction of Flavonoids from Olive (*Olea europaea*) Leaves, and Evaluation of Their Antioxidant and Anticancer Activities. *Molecules* **2018**, *23*, 2513. [CrossRef] [PubMed]
33. Chen, S.; Zeng, Z.; Hu, N.; Bai, B.; Wang, H.; Suo, Y. Simultaneous optimization of the ultrasound-assisted extraction for phenolic compounds content and antioxidant activity of *Lycium ruthenicum* Murr. fruit using response surface methodology. *Food Chem.* **2018**, *242*, 1–8. [CrossRef] [PubMed]

34. Ghosh, P.; Chandra-Pradhan, R.; Mishra, S.; Rout, P.K. Quantification and Concentration of Anthocyanidin from Indian Blackberry (*Jamun*) by Combination of Ultra- and Nano-filtrations. *Food Bioprocess Technol.* **2018**, *11*, 2194–2203. [CrossRef]

35. Tardugno, R.; Pozzebon, M.; Beggio, M.; Del Turco, P.; Pojana, G. Polyphenolic profile of *Cichorium intybus* L. endemic varieties from the Veneto region of Italy. *Food Chem.* **2018**, *266*, 175–182. [CrossRef] [PubMed]

36. Friščić, M.; Bucar, F.; Hazler-Pilepic, K. LC-PDA-ESI-MSn analysis of phenolic and iridoid compounds from *Globularia* spp. *J. Mass Spectrom.* **2016**, *51*, 1211–1236. [CrossRef] [PubMed]

37. Singleton, V.L.; Orthofer, R.; Lamuela-Raventós, R.M. Analysis of total phenols and other oxidation substrates and antioxidants by means of folin-ciocalteu reagent. *Methods Enzymol.* **1999**, *299*, 152–178.

38. Covarrubias-Cárdenas, A.G.; Martínez-Castillo, J.I.; Medina-Torres, N.; Ayora-Talavera, T.; Espinosa-Andrews, H.; García-Cruz, N.U.; Pacheco, N. Antioxidant Capacity and UPLC-PDA ESI-MS Phenolic Profile of *Stevia rebaudiana* Dry Powder Extracts Obtained by Ultrasound Assisted Extraction. *Agronomy* **2018**, *8*, 170. [CrossRef]

39. Jakobek, L.; Boc, M.; Barron, A.R. Optimization of Ultrasonic-Assisted Extraction of Phenolic Compounds from Apples. *Food Anal. Methods* **2015**, *8*, 2612–2625. [CrossRef]

40. Haya, S.; Bentahar, F.; Trari, M. Optimization of polyphenols extraction from orange peel. *J. Food Meas. Charact.* **2019**, *13*, 614–621. [CrossRef]

41. Ferreira, S.L.C.; Bruns, R.E.; Ferreira, H.S.; Matos, G.D.; David, J.M.; Brandao, G.C.; da Silva, E.G.P.; Portugal, L.A.; dos Reis, P.S.; Souza, A.S.; et al. Box-Behnken design: An alternative for the optimization of analytical methods. *Anal. Chim. Acta* **2007**, *597*, 179–186. [CrossRef] [PubMed]

42. Tan, M.C.; Chin, N.L.; Yusof, Y.A. A Box-Behnken Design for Determining the Optimum Experimental Condition of Cake Batter Mixing. *Food Bioprocess Technol.* **2012**, *5*, 972–982. [CrossRef]

43. He, Q.; Du, B.; Xu, B. Extraction Optimization of Phenolics and Antioxidants from Black Goji Berry by Accelerated Solvent Extractor Using Response Surface Methodology. *Appl. Sci.* **2018**, *8*, 1905. [CrossRef]

44. Paucar-Menacho, L.M.; Martínez-Villaluenga, C.; Dueñas, M.; Frias, J.; Peñas, E. Response surface optimization of germination conditions to improve the accumulation of bioactive compounds and the antioxidant activity in quinoa. *Int. J. Food Sci. Technol.* **2018**, *53*, 516–524. [CrossRef]

45. Zhou, J.; Zhang, L.; Li, Q.; Jin, W.; Chen, W.; Han, J.; Zhang, Y. Simultaneous Optimization for Ultrasound-Assisted Extraction and Antioxidant Activity of Flavonoids from *Sophora flavescens* Using Response Surface Methodology. *Molecules* **2019**, *24*, 112. [CrossRef] [PubMed]

46. Machado, A.P.D.F.; Pereira, A.L.D.; Barbero, G.F.; Martínez, J. Recovery of anthocyanins from residues of *Rubus fruticosus*, *Vaccinium myrtillus* and *Eugenia brasiliensis* by ultrasound assisted extraction, pressurized liquid extraction and their combination. *Food Chem.* **2017**, *231*, 1–10. [CrossRef]

47. Ruenroengklin, N.; Zhong, J.; Duan, X.; Yang, B.; Li, J.; Jiang, Y. Effects of Various Temperatures and pH Values on the Extraction Yield of Phenolics from Litchi Fruit Pericarp Tissue and the Antioxidant Activity of the Extracted Anthocyanins. *Int. J. Mol. Sci.* **2008**, *9*, 1333–1341. [CrossRef]

48. Mustafa, A.; Turner, C. Pressurized liquid extraction as a green approach in food and herbal plants extraction: A review. *Anal. Chim. Acta* **2011**, *703*, 8–18. [CrossRef] [PubMed]

49. Wijngaard, H.; Hossain, M.B.; Rai, D.K.; Brunton, N. Techniques to extract bioactive compounds from food by-products of plant origin. *Food Res. Int.* **2012**, *46*, 505–513. [CrossRef]

50. Heffels, P.; Weber, F.; Schieber, A. Influence of Accelerated Solvent Extraction and Ultrasound-Assisted Extraction on the Anthocyanin Profile of Different *Vaccinium* Species in the Context of Statistical Models for Authentication. *J. Agric. Food Chem.* **2015**, *63*, 7532–7538. [CrossRef]

51. Cai, Z.; Qu, Z.; Lan, Y.; Zhao, S.; Ma, X.; Wan, Q.; Jing, P.; Li, P. Conventional, ultrasound-assisted, and accelerated-solvent extractions of anthocyanins from purple sweet potatoes. *Food Chem.* **2016**, *197*, 266–272. [CrossRef] [PubMed]

52. Setyaningsih, W.; Saputro, I.E.; Palma, M.; Barroso, C.G. Stability of 40 Phenolic Compounds during Ultrasound-Assisted Extraction (UAE). *AIP Conf. Proc.* **2016**, *1755*, 080009. [CrossRef]

53. Machado, A.P.D.F.; Pasquel-Reátegui, J.L.; Barbero, G.F.; Martínez, J. Pressurized liquid extraction of bioactive compounds from blackberry (*Rubus fruticosus* L.) residues: A comparison with conventional methods. *Food Res. Int.* **2015**, *77*, 675–683. [CrossRef]

54. Carrera, C.; Ruiz-Rodríguez, A.; Palma, M.; Barroso, C.G. Ultrasound assisted extraction of phenolic compounds from grapes. *Anal. Chim. Acta* **2012**, *732*, 100–104. [CrossRef] [PubMed]

55. Cui, F.J.; Qian, L.S.; Sun, W.J.; Zhang, J.S.; Yang, Y.; Li, N.; Zhuang, H.N.; We, D. Ultrasound-Assisted Extraction of Polysaccharides from *Volvariella volvacea*: Process Optimization and Structural Characterization. *Molecules* **2018**, *23*, 1706. [CrossRef] [PubMed]

56. Rodríguez, K.; Ah-Hen, K.S.; Vega-Gálvez, A.; Vásquez, V.; Quispe-Fuentes, I.; Rojas, P.; Lemus-Mondaca, R. Changes in bioactive components and antioxidant capacity of maqui, *Aristotelia chilensis* [Mol] Stuntz, berries during drying. *LWT Food Sci. Technol.* **2016**, *65*, 537–542. [CrossRef]

57. AOAC. AOAC Peer Verified Methods Program. In *Manual on Policies and Procedures*; AOAC International: Rockville, MD, USA, 1998.

58. Fracassetti, D.; Del Bo, C.; Simonetti, P.; Gardana, C.; Klimis-Zacas, D.; Ciappellano, S. Effect of Time and Storage Temperature on Anthocyanin Decay and Antioxidant Activity in Wild Blueberry (*Vaccinium angustifolium*) Powder. *J. Agric. Food Chem.* **2013**, *61*, 2999–3005. [CrossRef] [PubMed]

59. Sui, X.; Bary, S.; Zhou, W. Changes in the color, chemical stability and antioxidant capacity of thermally treated anthocyanin aqueous solution over storage. *Food Chem.* **2016**, *192*, 515–524. [CrossRef] [PubMed]

60. Andrade, S.C.; Guiné, R.P.F.; Gonçalves, F.J.A. Evaluation of phenolic compounds, antioxidant activity and bioaccessibility in white crowberry (*Corema album*). *J. Food Meas. Charact.* **2017**, *11*, 1936–1946. [CrossRef]

61. Vázquez-Espinosa, M.; Espada-Bellido, E.; González de Peredo, A.V.; Ferreiro-González, M.; Carrera, C.; Palma, M.; Barroso, C.G.; Barbero, G.F. Optimization of Microwave-Assisted Extraction for the Recovery of Bioactive Compounds from the Chilean Superfruit (*Aristotelia chilensis* (Mol.) Stuntz). *Agronomy* **2018**, *8*, 240. [CrossRef]

62. Albuquerque, B.R.; Prieto, M.A.; Vazquez, J.A.; Barreiro, M.F.; Barros, L.; Ferreira, I.C.F.R. Recovery of bioactive compounds from *Arbutus unedo* L. fruits: Comparative optimization study of maceration/microwave/ultrasound extraction techniques. *Food Res. Int.* **2018**, *109*, 455–471. [CrossRef]

63. Nayak, B.; Dahmoune, F.; Moussi, K.; Remini, H.; Dairi, S.; Aoun, O.; Khodir, M. Comparison of microwave, ultrasound and accelerate-assisted solvent extraction for recovery of polyphenols from *Citrus sinensis* peels. *Food Chem.* **2015**, *187*, 507–516. [CrossRef]

64. Verbeyst, L.; Van Crombruggen, K.; Van der Plancken, I.; Hendrickx, M.; Van Loey, A. Anthocyanin degradation kinetics during thermal and high-pressure treatments of raspberries. *J. Food Eng.* **2011**, *105*, 513–521. [CrossRef]

65. Corrales, M.; Butz, P.; Tauscher, B. Anthocyanin condensation reactions under high hydrostatic pressure. *Food Chem.* **2008**, *110*, 627–635. [CrossRef]

66. Kovacevic, D.B.; Maras, M.; Barba, F.J.; Granato, D.; Roohinejad, S.; Mallikarjunan, K.; Montesano, D.; Lorenzo, J.M.; Putnik, P. Innovative technologies for the recovery of phytochemicals from *Stevia rebaudiana* Bertoni leaves: A review. *Food Chem.* **2018**, *268*, 513–521. [CrossRef]

67. Chemat, F.; Rombaut, N.; Maullemiestre, A.; Turk, M.; Perino, S.; Fabiano-Tixier, A.S.; Abert-Vian, M. Review of Green Food Processing techniques. Preservation, transformation, and extraction. *Innov. Food Sci. Emerg. Technol.* **2017**, *41*, 357–377. [CrossRef]

agronomy

MDPI

Article

Influence of Type of Management and Climatic Conditions on Productive Behavior, Oenological Potential, and Soil Characteristics of a 'Cabernet Sauvignon' Vineyard

Gastón Gutiérrez-Gamboa [1], Nicolás Verdugo-Vásquez [2] and Irina Díaz-Gálvez [3],*

[1]   Centro Tecnológico de la Vid y el Vino, Facultad de Ciencias Agrarias, Universidad de Talca,
       Av. Lircay S/N, Talca 3460000, Chile; ggutierrezg@utalca.cl
[2]   Instituto de Investigaciones Agropecuarias, INIA Intihuasi, Colina San Joaquín s/n, P.O. Box 36-B,
       La Serena 1700000, Chile; nicolas.verdugo@inia.cl
[3]   Instituto de Investigaciones Agropecuarias, INIA Raihuén, Casilla 34, San Javier 3660000, Chile
*   Correspondence: idiaz@inia.cl; Tel.: +56-9-32450885

Received: 30 November 2018; Accepted: 11 January 2019; Published: 1 February 2019

Abstract: (1) Background: Degradation of soils and erosion have been described for most of the soils presented along the Maule Valley. Organic and integrated management promotes agroecosystem health, improving soil biological activity. Due to this, the aim of this research was to study the effect of organic, integrated, and conventional management on the productive, oenological and soil variables of a vineyard cultivated under semiarid conditions during 5 consecutive seasons; (2) Methods: Yield, grape and wine oenological, and soil physicochemical parameters were evaluated. Bioclimatic indices were calculated in the studied seasons; (3) Results: Conventional management allowed to improve yield and the number of bunches per vine compared to organic management. However, this latter enhanced mineral nitrogen and potassium content in soil. Based on bioclimatic indices, heat accumulation improved number of bunches per plant and most of the soil physicochemical parameters; (4) Conclusions: Organic management improved the accumulation of some microelements in soils at the expense of yield. Organic matter decreased along the study was carried out. Season was the conditioning factor of the variability of most of the studied parameters, while the interaction between season and type of management affected soluble solids, probable alcohol and pH in grapes, and total polyphenol index and pH in wines.

Keywords: berry composition; Cabernet Sauvignon; conventional; integrated; organic; yield parameters

---

1. Introduction

Rainfed of the Maule Valley (Chile) has been characterized by a cereal, livestock, and viticulture agricultural history, all widely extractive activities [1]. Respect to the edaphic conditions, the soils from the rainfed areas of the Maule Valley presents low levels of organic matter (less than 2%) and low macro and micro nutrients content, mainly of boron [2]. Due to the extensive activities performed along this area, the soils present serious problems of erosion, and 80% exhibit the condition of non-arable, addition, rainfall does not amount to 600 mm·year^{-1}, and is concentrated from autumn to winter (May to August) [1]. Based on the aforementioned, together with the change in the use of agricultural land towards forestry activity, has led to small landowners to sell land and migrate to the city, generating important socio-cultural impacts [3].

Conventional viticulture disposes of agrochemicals such as inorganic fertilizers and synthetical chemical pesticides to the farming management, whereas these aforesaid products are banned in organic viticulture and only organic fertilizers and a certain non-synthetic pesticides are allowed in the agricultural management [4,5]. In organic viticulture, diseases control is mainly carried out using copper or sulfur treatment, while weed management is performed by tillage or grass-cutting [4,6,7]. Due to the high ability to adapt to the environment and its hardly manageable, weeds could affect productivity by competing with the vines [8]. The use of wheat and rye as cover crops may decrease weed growth, biomass, and density, without negative impacts on the vines [9,10]. In this way, the utilization of cover crops in the vineyards cultivated in rainfed conditions and in Mediterranean regions is not straightforward since water is the limiting factor and there is a severe competition for this resource [11]. Lower pruning weights and shoot length, fewer lateral shoots, and higher canopy openness have been observed in cover cropped vineyards [9,12]. However, long-term cover cropping using clover, ryegrass, and fescue is restricted in areas where water availability is limited [13]. It has been reported that, in these conditions, grapevines can be adapted, and a compensatory growth of their root system was observed, partly preventing the direct competition for resources between them and the cover crops [14]. In this scenario, the use of cover crops may avoid the dramatic reductions of stomatal conductance which occur in mid-summer and also allows to decrease yield and slightly increase grape quality [11]. These authors suggested that summer senescent and self-seeding herbaceous cover crops help to decrease soil erosion, which could improve organic matter and micro and macro nutrients content in soil.

Climate conditions had a strong influence on grape productivity delaying ripening and conditioning grape quality [15,16]. Certain bioclimatic indices have been developed in order to differentiate, describe and delimit distinct wine-growing sites. The Heliothermal Index (HI) uses daily temperatures during the period of the day in which grape metabolism is more active and includes a correction for length of the day in the case of higher latitude sites [17]. The Winkler Index (WI) was developed on the basis of the growing degree days, summed over the season on average, which allows to classify climates and to identify the grape varieties that can be fit to those regions [18]. The Biologically Effective Degree-Day Index (BEDD) allows to classify cold sites with low or late maturity potential, and hot sites with high or earlier maturity potential [19]. Night thermal conditions are associated to the accumulation of secondary metabolisms and can be estimated using the Cold Night Index (CI) [20]. In addition, bioclimatic indices associated with thermal accumulation such as mean thermal amplitude (MTA), average mean temperature of the warmest month (MTWM) and maximum average temperature of the warmest month (MATWM) can provide valuable information to characterize and compare different mesoclimates [21].

Based on the aforementioned, the aim of this work was to study the effect of organic, integrated and conventional management on the productive behavior, grape and wine oenological parameters, and soil physicochemical composition of a Cabernet Sauvignon vineyard located in Cauquenes, Maule Valley (Chile) during 5 consecutive seasons.

2. Materials and Methods

2.1. Study Site and Plant Material

The field trial was performed during 5 consecutive seasons (2003–2004, 2004–2005, 2005–2006, 2006–2007, and 2007–2008) in an experimental vineyard located into a rainfed area of Cauquenes, Maule Valley, Chile (35°58′ S, 72°17′ W; 177 meters above sea level). This area presents a sub humid Mediterranean climate. Cabernet Sauvignon (*Vitis vinifera* L.) ungrafted grapevines were planted in 1995 with a row orientation of east–west, trained to a double crossarms arrangements trellis system and pruned to a Guyot system, leaving about 40 buds per vine. The soil is Alfisol of granitic origin, slightly deep, of loamy clay texture horizon Ap Sandy loam with apparent density of 1.17 and 33% porosity, horizon Bt1, Bt2, and Bt3 clay with apparent density of 1.79, 1.75 and 1.77 (USDA). Plant density was 1428 plants·ha^{-1} with grapevine spacing between rows and within the row of 3.50 m × 2.00 m. Total surface planted was 2.37 ha. The vineyard was equipped with a drip irrigation system using 8 L·h^{-1}

drippers, to assure plant water needs when water is available. The vines were differentially irrigated from October to March when the leaf water potential reached 1.0 to 1.2 MPa.

2.2. Types of Managements

Three treatments or types of managements were performed in the vineyard. Organic management was carried out according to the Organic Industry Standards and Certification Committee [22]. Briefly, between rows were cover with green manure (*Avena sativa* and *Vicia faba*) in a dosage of 120 and 80 kg·ha^{-1}, while within rows were cover with different crops such as (i) a mixture of clover, (ii) *Medicago polymorpha* L., and (iii) *Lollium perenne* at a dosage of of 7.0, 7.0 and 3.0 kg·ha^{-1}, respectively. At sowing, 600 kg·ha^{-1} of phosphate rock was applied. After pruning, vine-shoots were crushed and incorporated to the soil. Additionally, compost of self-development skins, stalk and wheat straw were added at a dosage of 10 t·ha^{-1} at 20 cm of depth which equivale to 85 kg N·ha^{-1} per year. *Brevipalpus chilensis* management was performed using a water-miscible mineral oil preparation at a dosage of 1.5% applied at beginning of budburst. Subsequently, *Typhlodromus pyri* were released as biological control agent to the vineyard against *B. chilensis* at a relation of 1 to 6. Powdery mildew (*Uncinula necator*) management was performed using 20 kg·ha^{-1} of sulfur (powder) when the climate conditions were optimum for development for the disease. This was monitored from shoot development until veraison.

Integrated management was carried out according to the procedures of the International Organization for Biological and Integrated Control [23]. In brief, at the end of winter, Roundup was applied within rows at a dosage of 2 L·ha^{-1}. Natural weeds were maintained between rows. Monitoring and application of miscible mineral oil at 2% was carried out to *B. chilensis* control. After pruning, vine-shoots were crushed and incorporated to the soil. Sulfur application were made every 20 days from budburst to veraison to avoid grapevine diseases.

Conventional management in the vineyard was performed according to the exposed by Sotomayor et al. [24]. Briefly, conventional tillage was used to soil management, which remained without vegetation. Roundup was applied within rows at a dosage of 3 L·ha^{-1}. To avoid grapevine diseases, sulfur powder was applied at a dose of 20 kg·ha^{-1} every 15 days until veraison and a water-miscible mineral oil at 2% of Cyhexatin was applied at the beginning of the season and every 15 days until veraison for the control of *Brevipalpus chilensis*.

In all the vineyards, fertilization corresponded to the application of 100 kg·ha^{-1} of Urea, 100 kg·ha^{-1} of Triple Superphosphate (P_2O_5), 75 kg·ha^{-1} of Potassium Miurate (K_2O_5), 50 kg·ha^{-1} of Calcite Borate and 20 kg·ha^{-1} of Zinc Sulfate.

A randomized block design was performed into the vineyard, accounting 4 blocks, 3 treatments, and 3 replications. The sampling units corresponded to 21 plants, which were previously marked for identification.

2.3. Climatic and Soil Information

Climatic information was recorded since 2003 by a weather stations provided by Agromet [25] located around of 5 km from the site. Bioclimatic indices such as Huglin's Heliothermal Index (HI), Cool Night Index (CI), average mean temperature of the warmest month (MTWM) and maximum average temperature of the warmest month (MATWM) were calculated according to those mentioned by Gutiérrez-Gamboa et al. [26]. Briefly, HI was calculated from September first to March 31 by each season. This period corresponds to the growing stage of the Cabernet Sauvignon grapevines cultivated in Maule Valley, Chile. CI was calculated for March, while MTWM and MATWM were calculated for January. Mean thermal amplitude (MTA) corresponds to the sum of the differences between maximum and minimum daily temperature for March and was calculated according to the exposed by Tonietto and Carbonneau [20]. The Winkler Index (WI) was calculated from September first to March 31 for each season according to the exposed by Winkler et al. [18]. A pit was made in order to study soil physicochemical properties in the vineyard. pH, organic matter, nitrogen (N), phosphorus

(P), potassium (K), and boron (B) content were analyzed according to the methodology exposed by Sadzawka et al. [27].

2.4. Climatic Description of the Seasons

Precipitations calculated during each growing season (September to March) varied from 59.4 to 147.6 mm (Season 5 (S5) and Season 2 (S2), respectively). Bioclimatic indices are one of the common ways to characterize different climates based on the heat accumulation, vine growth and maturation potential [21]. Results of bioclimatic indices are shown in Table 1. Cool night index (CI) is a thermal indicator of night time temperature conditions at the end of the ripening stage, which is associated to the accumulation of secondary metabolites [20,26,28]. CI ranged from 9.4 to 12.0 (S3 and S4, respectively), with an average of 10.7. Huglin's heliothermal index (HI) is another bioclimatic index based on heat accumulation, which considers daily average temperature and daily maximum temperature, adjusting for day length during the season [20,29]. HI ranged from 2327.2 to 2416.4 (S2 and S5, respectively), with an average of 2354.9. The Winkler Index (WI) is calculated as the daily average temperature above a physiological base value of 10 °C. This is the minimum temperature at which grapevine growth occurs, for each day along its ripening from budburst through harvest stage [20,29,30]. WI ranged from 1559.6 to 1610.3 (S3 and S4, respectively), with an average of 1581.4. Mean thermal amplitude (MTA) reflects the differences between the maximum and minimum temperature at the end of the ripening stage [20]. MTA ranged from 439.5 to 522.4 (S1 and S5, respectively), with an average of 477.1. Certain bioclimatic indices such as the average mean temperature of the warmest month (MTWM) and the maximum average temperature of warmest month (MATWM) could give additional information about climate [26,28,31]. MTWM ranged from 20.8 to 22.2 °C (S2 and S5, respectively), with an average of 21.3 °C, while MATWM varied from 29.0 to 30.5 °C (S2-S4 and S5, respectively), with an average of 29.6 °C. Based on the aforementioned bioclimatic indices it is possible to characterize this climate as temperate.

Table 1. Information about bioclimatic indices such as cool night index (CI), Huglin's Heliothermal Index (HI), mean thermal amplitude (MTA), Winkler Index (WI), average mean temperature of the warmest month (MTWM), and maximum average temperature of the warmest month (MATWM) calculated for each study season, precipitations from September to March (Pp: mm) and harvest date.

Season	CI	HI	WI	MTA	MTWM	MATWM	Pp	Harvest Date
2003–2004 (S1)	11.0	2355.0	1577.4	439.5	21.3	29.8	105.4	March 23
2004–2005 (S2)	11.2	2327.2	1566.7	452.1	20.8	29.0	147.6	March 29
2005–2006 (S3)	9.4	2348.2	1559.6	514.7	21.1	29.5	112.0	March 21
2006–2007 (S4)	12.0	2377.5	1610.3	456.6	21.0	29.0	74.0	March 20
2007–2008 (S5)	10.2	2416.4	1593.0	522.4	22.2	30.5	59.4	March 25
Mean	10.7	2364.9	1581.4	477.1	21.3	29.6	99.7	-
SD	0.99	33.9	20.5	38.5	0.54	0.65	34.5	-

2.5. Productive and Oenological Parameters

Grapes had a good sanitary condition so it was processed immediately after the evaluations carried out. Harvest was performed when grapes reached 23–24 °Brix. Grapevine yield (kg·plant^{-1}) was measured using a precision balance (Sartorius, Gottingen, Germany). The number of bunches harvested by each grapevine was counted (N°bunches·plant^{-1}) and the weight of the bunches (g) was evaluated using the same precision balance with the aim to calculate the vineyard yield (kg·ha^{-1}). Weight of 100 berries (g) was also evaluated for to get an average (g berries^{-1}). Ravaz index was calculated according to the exposed by Shellie [32], using yield and pruning weight.

Soluble solids (°Brix), probable alcohol, total acidity (g·L^{-1} of sulfuric acid) and pH were measured according to the International Organization of Vine and Wine (OIV) methodologies [33]. Total polyphenols index was determined in wines based on the stated by Amerine et al. [34].

2.6. Statistical Analysis

A general linear model, using the statistical program Statgraphics Centurion XVI.I (Virginia, USA) was performed. The treatments (types of management) were considered as a fixed variable, while the season (associated with climatic conditions) was considered as a random variable in the general linear model. Differences between samples were compared using the Tukey test at 95% of probability level. Principal component analysis (PCA) was performed using each variable together with the bioclimatic indices calculated from field data for the site, through InfoStat software (www.infostat.com.ar).

3. Results and Discussion

3.1. Grapevine Productivity

The effect of type of management (organic, integrated and conventional) evaluated during 5 consecutive seasons on yield (t·ha^{-1}), Ravaz index and yield components such as number of bunches per plant, weight of bunches (g), weight of berries (g), and number of berries per bunch is shown in Table 2. Weight of bunches, weight of berries, and number of berries per bunch were not affected by the types of managements, except for the number of bunches per plant, which varied from 34 to 42. This parameter together with yield was higher in the conventional than the organic management in the vineyard. Number of berries per bunch was more influenced by season than the type of management and their interaction (Figure S1). Yield ranged from 5.95 to 4.83 (t·ha^{-1}). Ravaz index varied from 2.96 to 2.05 and was higher in the conventional than the organic management, and this parameter was more influenced by season than the type of management and their interaction (Figure S1). Ravaz index represents the reproductive to vegetative growth ratio [35]. Balanced vines present Ravaz index values from 3 to 10, with optimal values between 5 to 7 [36]. The Ravaz Index balanced to 3, indicating an excess of vigor at the expense of yield [35]. Based on this, the vines managed organically present greater vigor than conventional management and this could indicate that the productive potential of these vines could be higher. This growth can be detrimental to the regulation of the bud differentiation, allowing a worse cluster microclimate, negatively affecting number of bunches per plant [37]. However, the aforesaid differences were mainly associated to climate conditions reached in each season (Table 2). Different results are reported in literature however, most of them have reported that organic management negatively affected yield in grapevine. Malusà et al. [38] showed that yield in organic vineyards were lower than in the vineyards under conventional management. Döring et al. [39] reported that the grapevines growing under organic management showed significantly lower growth and yield in comparison to the integrated treatment, which was associated with differences in growth and cluster weight. Brunetto et al. [40] reported that in a first study season, grapevines under organic management showed higher yield than the conventional ones. However, no differences were found in yield in the second season Pou et al. [11] reported that a mixture of perennial grasses and legumes as cover crops increased Ravaz index compared to no tillage and traditional tillage, offering a better balance between vegetative and reproductive growth. However, the opposite was observed by these authors, when yield reduction in cover cropped vines was the highest and counteracted their lower pruning weight. Season influenced all the productivity parameters of the grapevines (Table 2). Yield and Ravaz index reached in the second season were higher than the found in the first and third seasons, while weight of bunches and number of berries per bunch reached in the second season were the highest. Additionally, this season reached the lowest number of bunches per plant. Based on this, it is possible that there was some compensation effect on yield component among the seasons, as has been discussed by Sadras et al. [41]. The third season presented a higher weight of berries than the fifth season.

To classify the different types of managements by season according to grapevine productivity, principal component analysis (PCA) was carried out using yield parameters obtained from organic (O), integrated (I), and conventional (C) managements, together with different bioclimatic indices, such as cool night index (CI), Huglin's Heliothermal Index (HI), mean thermal amplitude (MTA),

Winkler Index (WI), average mean temperature of the warmest month (MTWM), and maximum average temperature of the warmest month (MATWM) performed in a Cabernet Sauvignon vineyard along five consecutive seasons (S). Principal component 1 (PC1) explained 40.2% of the variance and PC2 explained 23.8% of the variance, representing 64.0% of all variance (Figure 1). PC1 was strongly correlated with HI index, MTWM and MATWM, while PC2 was correlated with yield, Ravaz index and number of berries per bunch. Both components allowed to separate the different treatments by season. Second season was correlated with high amounts of weight of bunches and number of berries per bunch, while fifth season was correlated with most of the bioclimatic indices. Accumulated rainfall during the second season was the highest compared to the rest of the seasons. This might condition most of the productive measured parameters. Besides, HI and WI indices were positively correlated with number of bunches per plant and Ravaz index, while HI index was negatively correlated with weight of berries and weight of bunches. MATWM and MTWM were correlated negatively with weight of berries.

Based on this, bioclimatic indices associated with the accumulation of effective degree days, such as HI and WI were positively correlated with the number of bunches per plant, while bioclimatic indices associated with the synthesis of secondary metabolites as MATWM and MTWM were inversely correlated with the weight of berries per bunch. It is of wide knowledge that the number of bunches per grapevine is determined during the previous year. Light and temperature exposure have the greatest potential to regulate the differentiation of anlagen into either bunch or tendril primordia [42]. Based on this, high intensity of light and moderate temperatures on the bud after different viticultural managements favor the formation of cluster primordia [43,44]. The season factor influenced the number of bunches per plant more than the type of management. Number of bunches per plant was higher in the first, fourth, and fifth seasons than the in the rest of the study seasons, while the second season presented the lowest number. However in the only fourth season was presented a high heat accumulation based on WI, while the second season presented the lowest values of HI, MTWM, and MATWM. Therefore, data in relation to radiation can give more information to the obtained results. Respect to the weight of berries, it is probably that the low amount of this parameter reached in the warmer season is due to the dehydration of the berry, which was described by Bonada et al. [44]. None of the bioclimatic indices were related with yield, which was affected by the season.

Table 2. Effect of organic, integrated and conventional management on yield (t·ha^{-1}), Ravaz index, and yield components such as number of bunches per plant, weight bunches, weight of berries and number of berries per bunch in Cabernet Sauvignon grapevines evaluated during 5 consecutive seasons.

Factor	Yield (t·ha^{-1})	Ravaz Index	N°Bunches· Plant^{-1}	Weigh of Bunches (g)	Weight of Berries (g)	N°Berries· Bunch^{-1}
Type of management (M)						
Organic	4.8 a	2.1 a	33.7 a	81.1	0.98	83.8
Integrated	5.5 ab	2.4 ab	37.0 ab	82.2	0.98	84.1
Conventional	6.0 b	3.0 b	41.5 b	95.3	1.03	92.9
Season (S)						
S1	4.4 a	1.4 a	48 c	60.1 a	1.02 ab	58.7 a
S2	6.4 c	2.9 b	26 a	173.1 c	1.00 ab	174.2 c
S3	4.8 ab	1.5 a	32 b	83.3 b	1.12 b	75.3 ab
S4	5.7 bc	3.4 b	42 c	77.4 b	1.05 ab	72.5 ab
S5	5.9 bc	4.1 b	42 c	79.7 b	0.82 a	98.7 b
Significance (*p*-value)						
M	0.02	0.05	0.0076	0.054	0.62	0.54
S	0.004	0.0005	0.0001	0.00001	0.03	0.0004
M × S	0.6	0.007	0.57	0.22	0.06	0.01

M × S: Interaction between type of management (M) and season (S). For a given factor and significance $p \leq 0.05$, different letters within a column represent significant differences (Tukey's test, $p \leq 0.05$).

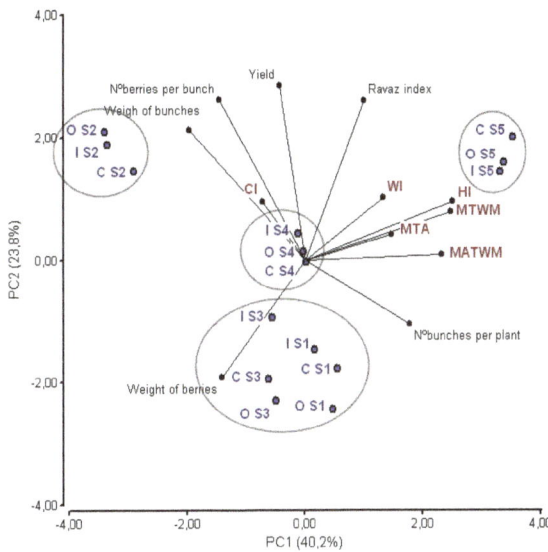

Figure 1. Principal component analysis (PCA) performed with grapevine productive parameter from a Cabernet Sauvignon vineyard planted in an organic (O), integrated (I), and conventional (C) managements, together with different bioclimatic indices such as cool night index (CI), Huglin's Heliothermal Index (HI), mean thermal amplitude (MTA), Winkler Index (WI), average mean temperature of the warmest month (MTWM), and maximum average temperature of the warmest month (MATWM) along five consecutive seasons (S).

3.2. Berry Composition

The effect of type of management (organic, integrated, and conventional) evaluated during 5 consecutive seasons on berry composition is shown in Table 3. None of the berry components were affected by the type of management, due to the treatments were harvested in the same stage of ripening. Little differences in berry composition have been reported by some authors in grapevines growing under changes on the types of managements. As was observed in Table 3, Döring et al. [39] reported that fruit quality parameters were not affected by the management system. These aforementioned results include the biodynamic system. Additionally, Tesic et al. [45] showed that only in the third study year, grapes from Chardonnay grapevines growing under organic management presented higher soluble solids and total acidity, together with a lower pH than the grapes from grapevines growing under conventional management. Soluble solids and probable alcohol were not affected by the season. However, total acidity, soluble solids to total acidity ratio, and pH were affected by the season factor. Total acidity ranged from 2.52 to 4.82 g·L^{-1} of sulfuric acid. Total acidity content found in the fourth season presented the highest total acidity content, while the second season reached the lowest amount.

To classify the different types of managements by season according to berry composition, a PCA was carried out using grape physiochemical parameters from grapevines under organic (O), integrated (I) and conventional (C) managements, together with the different bioclimatic indices, such as CI, HI, MTA, WI, MTWM and MATWM performed in a Cabernet Sauvignon vineyard along five consecutive seasons (S). PC1 explained 38.4% of the variance and PC2 explained 28.0% of the variance, representing 66.4% of all variance (Figure 2). PC1 was strongly correlated with HI index and MTWM, while PC2 was only strongly correlated with CI. Both components allowed separation of the treatments by season. The second season was correlated with soluble solids to total acidity ratio (SS to TA ratio). Third season was positively correlated with probable alcohol, °Brix and pH, while was negatively correlated with the fourth season and WI index. As in Figure 1, fifth season was correlated with most of the bioclimatic indices. Besides, CI and WI indices were inversely correlated with pH, while WI index was correlated with total acidity. Based on these results, high temperatures during the month before harvest allowed to reach a low pH in grapes. High night temperatures tend to promote vegetative growth at the expense of productivity [46]. HI index was correlated with total acidity and inversely correlated with SS to TA ratio. Based on this, the bioclimatic indices associated with the heat accumulation or growing degree-days such as HI and WI were related with high values of total acidity. These results observed in Figure 2 were, somehow, unexpected since heat accumulation and low temperatures at the end of ripening stage is related to a delay in grape maturation associated with low pH and high total acidity. However, these bioclimatic indices are calculated based on the active growth temperature of the vine, considering corrections by maximum and minimum temperatures, day–night fluctuations, and latitude. According to Figure 2, bioclimatic indices associated to warm conditions such as MTA, MTWM, and MATWM were more related to high pH and °Brix. pH as soluble solids and probable alcohol were affected by the interaction between the type of management and season (Figure S2). In this way, the accumulated rainfall during the S1 and S3 growing season were higher than to those obtained in S4, which showed warmer temperature conditions. Additionally, the grapevines cultivated under organic management showed lower Ravaz index than the conventional ones, which was associated to a high vigor [11,35]. Therefore, vineyard productivity, which was influenced by climatic conditions and type of management, considerably affected berry composition in terms of soluble solids, probable alcohol and pH.

Table 3. Effect of organic, integrated, and conventional management on berry components in Cabernet Sauvignon grapevines evaluated during 5 consecutive seasons.

Factor	Soluble Solids (°Brix)	Probable Alcohol	Total Acidity [a]	SS·TA^{-1} [b]	pH
Type of management (M)					
Organic	24.9	14.2	3.85	6.31	3.47
Integrated	25.1	14.3	3.90	6.44	3.47
Conventional	25.2	14.4	4.00	6.48	3.49
Season (S)					
S1	25.1	14.3	3.99 b	6.29 b	3.40 a
S2	25.2	14.4	2.52 a	10.05 c	3.53 bc
S3	24.9	14.2	4.25 b	5.87 b	3.57 c
S4	24.8	14.2	4.82 c	5.15 a	3.45 ab
S5	25.2	14.4	4.13 b	6.10 b	3.47 abc
Significance (*p*-value)					
M	0.17	0.20	0.16	0.41	0.51
S	0.45	0.56	0.0001	0.0001	0.0016
M × S	0.0001	0.00001	0.07	0.087	0.0007

M × S: Interaction between type of management (M) and season (S). SS·TA: Soluble solids (SS) to total acidity (TA) ratio. [a] As g·L^{-1} of sulfuric acid. [b] Soluble solid to total acidity ratio. For a given factor and significance $p \leq 0.05$, different letters within a column represent significant differences (Tukey's test, $p \leq 0.05$).

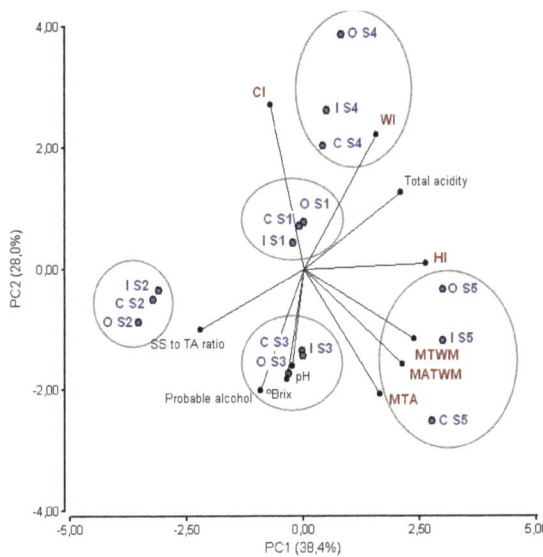

Figure 2. Principal component analysis (PCA) performed with berry composition from a Cabernet Sauvignon vineyard planted in an organic (O), integrated (I), and conventional (C) managements, together with different bioclimatic indices such as cool night index (CI), Huglin's Heliothermal Index (HI), mean thermal amplitude (MTA), Winkler Index (WI), average mean temperature of the warmest month (MTWM), and maximum average temperature of the warmest month (MATWM) along five consecutive seasons (S).

3.3. Wine Oenological Parameters

The effect of type of management (organic, integrated, and conventional) evaluated during 5 consecutive seasons on wine oenological parameters is shown in Table 4. Type of management did not affect the wine composition in terms of its alcohol degree and pH. However, the wines produced from grapevines planted in the integrated management presented higher total phenols than the organic wines. Provost and Pedneault [12] reviewed organic management vineyard and its impacts on wine quality. These authors suggested that organic management may increase the level of certain compounds in wines with little impacts on wine sensory perception. Martin and Rasmussen [47] indicated that organic managed grapevines suffer more levels of biotic stress than conventional grapevines and may probably produce higher rate of secondary metabolites, mainly phenolic compounds. These aforementioned results did not match with the shown in Table 4. Additionally, the wines from grapevines planted in the conventional management presented higher total acidity than the organic wines. Season had a considerable effect on wine oenological parameters. The wines elaborated in the first season presented the lowest content of total phenols and total acidity. The wines of the third and the fourth seasons reached lower alcoholic degree than the wines of the rest of the studied seasons. However, in all samples, this oenological parameter was high (>14% of alcoholic degree). The wines elaborated from the second season presented higher pH than the wines from the third and the fifth seasons.

To classify the different types of managements by season according to wine oenological parameters, a PCA was carried out using the measured physiochemical parameters in wines from grapevines under organic (O), integrated (I), and conventional (C) managements, together with the different bioclimatic indices, such as CI, HI, MTA, WI, MTWM, and MATWM performed in a Cabernet Sauvignon vineyard along five consecutive seasons (S). PC1 explained 42.8% of the variance and PC2 explained 30.3% of the variance, representing 73.1% of all variance (Figure 3). PC1 was strongly correlated with HI index and MTWM, while PC2 was strongly correlated with total phenolics and total acidity. Both components allowed to separate the treatments by season. Second and fourth seasons were correlated pH. Third season was positively correlated with total phenolics and total acidity. This season was inversely correlated with alcohol degree and the first season. Besides, all the bioclimatic indices except CI and WI were inversely correlated with pH. These results were unexpected since warm climate conditions and high temperatures at the end of ripening stage are mainly related to early ripening [26,28,31]. These authors showed that the sites that present warm climate conditions lead to wines with high alcohol degree and pH, together with low total acidity [26,28,31]. All the wine oenological parameters measured were influenced by the interaction season x type of management (Figure S3). Total polyphenol index and pH in wines were more influenced by the interaction between type of management and season, while alcohol degree and total acidity, including total polyphenol index, were also influenced by the season (Table 4). Canopy shade down regulate gene expression in the anthocyanin biosynthesis pathway [48]. Cluster shading decreased the accumulation of flavonols and skin proanthocyanidins with minimal differences in anthocyanins in pinot noir berries [49]. Organic management presented low values of Ravaz index, which is mainly associated with high vigor in vines (Table 2). Low vigor vines lead to higher wine anthocyanins, total phenols and color intensity than high vigor vines [50]. Moreover, S1 which reached the lowest total polyphenol index presented higher MATWM than the rest of the seasons with the exception of S5 (Table 1). The accumulation of flavanols and hydroxycinnamic acids was inversely related to MATWM in Carignan grapes [51].

Table 4. Effect of organic, integrated and conventional management on wine oenological parameters obtained from Cabernet Sauvignon grapevines evaluated during 5 consecutive seasons.

Factor	Total Polyphenol Index	Alcohol Degree (% v/v)	Total Acidity [a]	pH
Type of management (M)				
Organic	63.41 a	14.88	3.65 a	3.48
Integrated	80.94 b	15.28	3.82 ab	3.46
Conventional	73.68 ab	15.12	3.95 b	3.48
Season (S)				
S1	23.33 a	15.56 b	2.52 a	3.52 bc
S2	88.57 bc	15.36 b	4.25 c	3.56 c
S3	106.82 bc	14.26 a	4.82 d	3.44 ab
S4	106.8 c	14.54 a	4.13 c	3.47 abc
S5	78.34 b	15.66 b	3.48 b	3.38 a
Significance (*p*-value)				
M	0.027	0.068	0.0237	0.736
S	0.00001	0.0004	0.00001	0.002
M × S	0.00001	0.0105	0.011	0.00001

M × S: Interaction between type of management (M) and season (S). [a] As $g \cdot L^{-1}$ of sulfuric acid. For a given factor and significance $p \leq 0.05$, different letters within a column represent significant differences (Tukey's test, $p \leq 0.05$).

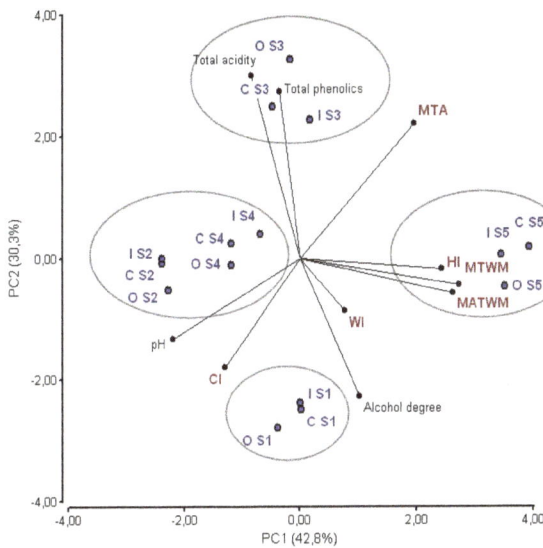

Figure 3. Principal component analysis (PCA) performed with wine oenological parameter obtained from Cabernet Sauvignon grapevines planted in an organic (O), integrated (I), and conventional (C) managements, together with different bioclimatic indices such as cool night index (CI), Huglin's Heliothermal Index (HI), mean thermal amplitude (MTA), Winkler Index (WI), average mean temperature of the warmest month (MTWM), and maximum average temperature of the warmest month (MATWM) along five consecutive seasons (S).

3.4. Soil Physicochemical Characteristics

The effect of type of management (organic, integrated, and conventional) evaluated during 5 seasons on soil physicochemical characteristics is shown in Table 5. Soils under the organic management showed the highest amount of N and K. N varied from 3.49 to 4.37, while K ranged from 165.7 to 185.3 (conventional and organic managements, respectively). The optimum mineral N content for grapevine development corresponds to the range of 15–25 ppm, while for K, the optimum level is 140 ppm [27]. Type of management did not affect pH, organic matter, P and B content in soils. pH ranged from 5.97 to 6.14 (integrated and conventional managements, respectively). These values are within the optimum development range of the grapevines [27]. However, there could be a tendency to increase the soil acidity, which is explained by the low content of organic matter reached [36]. Despite the aforementioned, organic matter was higher than to those exposed by Gutiérrez-Gamboa and Moreno-Simunovic [26] in rainfed vineyards from the Maule Valley.

These results differ to those reported by Fließbach et al. [52]. These authors showed that applying organic fertilization, the content of organic matter was improved compared to the traditional exploitation soils. Different results are reported in literature about the effect of type of management on soil chemical composition. In this way, Morlat and Chaussod [53] reported that long term additions of organic amendments improved soil water holding capacity and P and K content. In addition, Coll et al. [4] reported that organic management led to an increase in soil organic matter, P content, soil microbial biomass, plant feeding and fungal feeding nematode densities. These authors also reported that organic farming increased soil compaction, decreased endogenic earthworm density and not modified the soil micro food web evaluated by nematofauna analysis. Additionally, Steenwerth and Belina [54] reported that cover crops enhance soil organic matter, carbon dynamic and microbiological function in vineyards ecosystems. Type of management did not affect organic matter in soils, including the grapevines cultivated under organic management. It is probably that when organic matter is applied to a soil, it accumulates until reaching an equilibrium. This balance is achieved when the rate of mineralization (action of microorganisms) is equal to the rate of incorporation of organic matter, that is, the amount applied is equal to the amount that microorganisms use to develop their biological processes [55]. Tillage brings subsurface soil to the surface where it is then exposed to wet–dry and freeze–thaw cycles and subjected to raindrop impact [56]. This results in an increase in the susceptibility of aggregates to disruption [57]. Plowing changes soil physicochemical parameters, increasing the decomposition rates of litter [58]. The chemical and colloidal properties of soil organic matter can be studied only in the free state, that is, when it is separated from the inorganic compounds of the soil. Thus, the first task in research is to separate the organic matter from the inorganic matrix: from sand, silt and clay. In clay soils, the C content of macroaggregates was 1.65 times greater compared to microaggregates [59]. Due to the aforementioned, it is probably that there were no differences among the type of managements on organic matter on soils. Season was the most important factor of the variability of organic matter and its content tended to decrease over time according to the accumulated rainfall (Table 5). Temperature sensitivity values demonstrated a strong positive correlation with annual precipitation, so C decomposition in soils from zones with high precipitation exhibits increased temperature sensitivity [60]. Additionally, N contained in organic matter would cause an initial increase in pH, associated with the formation of NO_4^+ that consumes protons. The subsequent nitrification would result in a decrease in pH due to release of the protons to the soil solution. The decrease of the pH by formation of NO_3 would not achieve the original levels of acidity since a high concentration of NH_4^+ has a nitrification inhibiting effect [61].

P ranged from 6.83 to 7.89, while B varied from 1.23 to 1.38 (conventional and organic managements, respectively). With respect to the description by Sadzawka et al. [27], P should be higher than 8 ppm, and B between 1 and 2 ppm. According to the shown in Table 5, soils lack of mineral nitrogen. In this way, nitrogen accessibility by grapevines in rainfed conditions relies among other factors on the presence of sufficient soil water, which under Mediterranean climate conditions is mostly accumulated during winter and/or early spring rainfalls [28]. Season affected all the evaluated parameters with the exception of pH. During the fourth and fifth seasons the soil reached lower organic matter and N than in the rest of the study seasons. In the third season was showed the highest N and B content in soil. During the second seasons was showed the lowest P content, while in the fifth season was presented the lowest K content in soils.

To classify the different types of managements by season according to soil characteristics of the vineyard, a PCA was carried out using the measured parameters in soils from the vineyard under organic (O), integrated (I) and conventional (C) managements, together with the different bioclimatic indices, such as CI, HI, MTA, WI, MTWM, and MATWM performed along five consecutive seasons (S). PC1 explained 46.2% of the variance and PC2 explained 28.8% of the variance, representing 75.0% of all variance (Figure 4). PC1 was strongly correlated with organic matter, K, HI, MTWM, and MATWM, while PC2 was strongly correlated with B and CI. Both components allowed to separate the treatments by season. CI was negatively related with B. HI was negatively related with organic matter, N and K. MATWM was negatively related with K. MTA was positively related with B and negatively with K. MTWM was negatively related with N and K. WI was negatively related with organic matter, N and B. In this way, heat accumulation promotes low pH, organic matter, K and B in soil. Additionally, thermal amplitude was related with high B content in soil. It is probably that temperature affects decomposition organic matter of soil, affecting the rest of the measure parameters. The effects of temperature on soil organic matter decomposition have been reviewed by certain authors such as Kätterer et al. [62] and Conant et al. [63].

Table 5. Effect of organic, integrated and conventional management on soil physicochemical characteristics obtained from a Cabernet Sauvignon vineyard evaluated during 5 consecutive seasons.

Factor	pH	Organic Matter (%)	N (ppm)	P (ppm)	K (ppm)	B (ppm)
Type of management (M)						
Organic	6.07	1.51	4.37 b	7.29	185.3 b	1.38
Integrated	5.97	1.34	3.53 a	6.91	172.1 a	1.26
Conventional	6.14	1.45	3.49 a	6.83	165.7 a	1.23
Season (S)						
S1	6.07	1.73 b	4.62 b	6.60 ab	173.0 b	1.21 a
S2	6.39	2.07 b	4.54 b	6.39 a	194.1 c	1.09 a
S3	5.98	1.56 b	6.07 c	7.21 bc	180.1 bc	2.38 b
S4	6.05	1.06 a	2.83 a	7.43 bc	179.7 bc	0.91 a
S5	5.83	1.16 a	2.09 a	7.60 c	144.9 a	1.15 a
Significance (*p*-value)						
M	0.63	0.26	0.011	0.11	0.0008	0.25
S	0.22	0.0005	0.00001	0.0036	0.00001	0.00001
M × S	0.37	0.064	0.77	0.99	0.99	0.84

M × S: Interaction between type of management (M) and season (S). For a given factor and significance $p \leq 0.05$, different letters within a column represent significant differences (Tukey's test, $p \leq 0.05$).

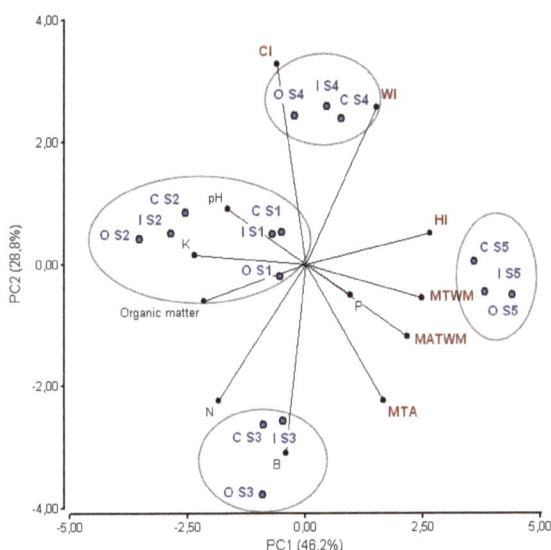

Figure 4. Principal component analysis (PCA) performed with soil physicochemical parameters obtained from Cabernet Sauvignon vineyard planted in an organic (O), integrated (I), and conventional (C) managements, together with different bioclimatic indices such as cool night index (CI), Huglin's Heliothermal Index (HI), mean thermal amplitude (MTA), Winkler Index (WI), average mean temperature of the warmest month (MTWM), and maximum average temperature of the warmest month (MATWM) along five consecutive seasons (S).

4. Conclusions

Type of management whether organic, conventional or integrated affected productive, wine oenological and soil physicochemical parameters. However, none of the grape oenological parameters measured were affected by the type of management. Conventional management showed higher yield, Ravaz index, number of bunches per vine and wine total acidity than the organic management.

However, organic management improved soil N and K content compared to conventional and integrated managements. Integrated management lead to higher total phenols than organic management. Season factor had mostly influenced productive parameters, grape oenological parameters with the exception of soluble solids, wine oenological parameters and soil chemical parameters except pH. Interaction between type of management and season influenced soluble solids, probable alcohol and pH in grapes, and total polyphenol index and pH in wines. Organic matter decreased along the study were carried out, being the season the most important factor of variability. Based on the bioclimatic indices, heat accumulation conditioned number of bunches per plant, leading to grapes and wines with low pH. In addition, heat accumulation also affected organic matter, pH and some micronutrients content in soils. In this way, thermal amplitude was positively related with B, which is a scarce microelement in rainfed soils from Maule Valley.

Supplementary Materials: The following are available online at http://www.mdpi.com/2073-4395/9/2/64/s1, Figure S1: Significant interactions of analysis of variance for productive variables; Figure S2: Significant interactions of analysis of variance for berry components; Figure S3: Significant interactions of analysis of variance for wine oenological parameters.

Author Contributions: Data analysis and wrote the paper: G.G.-G. Data analysis and critical revision: N.V.-V. Conceived and designed the experiment: I.D.-G.

Funding: This research was funded by the Cooperation Agreement of the Chilean Agriculture Ministry and the Swiss Agricultural Federal Office, within the framework of the project "Development and diffusion of new systems of organic and integrated production of grapes for the elaboration of wines in the Province of Cauquenes."

Acknowledgments: G.G.G. thanks for the financial support given by CONICYT, BCH/Doctorado-72170532. I.D.G. and N.V.V. thanks for the financial support given by sub research direction of INIA. Authors thank to the project "Aseguramiento de la sustentabilidad de la viticultura nacional frente a los nuevos escenarios que impone el cambio climático".

Conflicts of Interest: The authors declare no conflict of interest.

References

1. Ovalle, C.; del Pozo, A.; Avendaño, J.; Fernández, F.; Arredondo, S.; Aravena, T.; Cares, J.; Aronson, J.; Longeri, L.; Herrera, A. Restauración y rehabilitación de agroecosistemas degradados en el Secano Interior Mediterráneo de Chile. In *Recuperación y Manejo de Ecosistemas Degradados*; IICA: San Jose, Costa Rica, 1998; Volume 49, pp. 97–111.
2. Del Canto, S.P.; Del Pozo, L.A. *Áreas Agroclimáticas y Sistemas Productivos en la VII y VIII Regiones*; Instituto de Investigaciones Agropecuarias (INIA): Chillán, Chile, 1999; Volume 35.
3. Rojas, G.A. Heritage and Wine Identity. Reflections on the evolution of the cultural meanings of wine in Chile. *Rivar* **2015**, *2*, 88–105.
4. Coll, P.; Le Cadre, E.; Blanchart, E.; Hinsinger, P.; Villenave, C. Organic viticulture and soil quality: A long-term study in Southern France. *Appl. Soil Ecol.* **2011**, *50*, 37–44. [CrossRef]
5. Briar, S.S.; Grewal, P.S.; Somasekhar, N.; Stinner, D.; Miller, S.A. Soil nematode community, organic matter, microbial biomass and nitrogen dynamics in field plots transitioning from conventional to organic management. *Appl. Soil Ecol.* **2007**, *37*, 256–266. [CrossRef]
6. Kuflik, T.; Prodorutti, D.; Frizzi, A.; Gafni, Y.; Simon, S.; Pertot, I. Optimization of copper treatments in organic viticulture by using a web-based decision support system. *Comput. Electron. Agric.* **2009**, *68*, 36–43. [CrossRef]
7. Garde-Cerdán, T.; Mancini, V.; Carrasco-Quiroz, M.; Servili, A.; Gutiérrez-Gamboa, G.; Foglia, R.; Romanazzi, G. Chitosan and Laminarin as Alternatives to Copper for Plasmopara viticola Control: Effect on Grape Amino Acid. *J. Agric. Food Chem.* **2017**, *65*, 7379–7386. [CrossRef] [PubMed]
8. Wisler, G.C.; Norris, R.F. Interactions between weeds and cultivated plants as related to management of plants as related to management of plant pathogens. *Weed Sci.* **2005**, *53*, 914–917. [CrossRef]
9. Krohn, N.G.; Ferree, D.C. Effects of low-growing perennial ornamental groundcovers on the growth and fruiting of seyval blanc grapevines. *HortScience* **2005**, *40*, 561–568.

10. Valdés-Gómez, H.; Gary, C.; Cartolaro, P.; Lolas-Caneo, M.; Calonnec, A. Powdery mildew development is positively influenced by grapevine vegetative growth induced by different soil management strategies. *Crop. Prot.* **2011**, *30*, 1168–1177. [CrossRef]

11. Pou, A.; Gulías, J.; Moreno, M.; Tomàs, M.; Medrano, H.; Cifre, J. Cover cropping in *Vitis vinifera* L. cv. Manto Negro vineyards under Mediterranean conditions: Effects on plant vigour, yield and grape quality. *OENO One* **2011**, *45*, 223–234. [CrossRef]

12. Provost, C.; Pedneault, K. The organic vineyard as a balanced ecosystem: Improved organic grape management and impacts on wine quality. *Sci. Hortic.* **2016**, *208*, 43–56. [CrossRef]

13. Guerra, B.; Steenwerth, K. Influence of floor management technique on grapevine growth, disease pressure, and juice and wine composition: A review. *Am. J. Enol. Vitic.* **2012**, *63*, 149–164. [CrossRef]

14. Celette, F.; Gaudin, R.; Gary, C. Spatial and temporal changes to the water regime of a Mediterranean vineyard due to the adoption of cover cropping. *Eur. J. Agron.* **2008**, *29*, 153–162. [CrossRef]

15. Van Leeuwen, C.; Seguin, G. The concept of terroir in viticulture. *J. Wine Res.* **2006**, *17*, 1–10. [CrossRef]

16. Gutiérrez-Gamboa, G.; Moreno-Simunovic, Y. Location effects on ripening and grape phenolic composition of eight 'Carignan' vineyards from Maule Valley (Chile). *Chil. J. Agric. Res.* **2018**, *78*, 139–149. [CrossRef]

17. Huglin, P. Nouveau mode d'évaluation des possibilités héliothermiques d'un milieu viticole. *C. R. Acad. Agric. Fr.* **1978**, *64*, 1117–1126.

18. Winkler, A.J.; Cook, J.A.; Kliewer, W.M.; Lider, L.A. *General Viticulture*; University of California Press: Berkeley, CA, USA, 1974; ISBN 978052005912.

19. Gladstones, J. *Viticulture and Environment*; Winetitles: Adelaide, Australia, 1992; 310p, ISBN 1875130128.

20. Tonietto, J.; Carbonneau, A. A multicriteria climatic classification system for grape-growing regions worldwide. *Agric. For. Meteorol.* **2004**, *124*, 81–97. [CrossRef]

21. Jones, G.V.; Duff, A.A.; Hall, A.; Myers, J.W. Spatial analysis of climate in winegrape growing regions in the western United States. *Am. J. Enol. Vitic.* **2010**, *61*, 313–326.

22. OISCC. National Standard for Organic and Biodynamic Produce, Edition 3.5, Last Updated 1 February 2013. Canberra: Organic Industry Standards and Certification Committee. Available online: http://www.daff.gov.au/__data/assets/pdf_file/0018/126261/nationalstandard-2013.pdf (accessed on 10 December 2018).

23. IOBC. Directrices para la Producción Integrada de uva. 1999. Available online: https://www.iobc-wprs.org/ip_ipm/IOBC_Guideline_Grapes_1999_ESPANOL.pdf (accessed on 10 December 2018).

24. Sotomayor, J. Establecimiento y manejo de viñedos modernos en el secano interior centro sur de Chile. In *Proposiciones Tecnológicas para un Desarrollo Sustentable del Secano*; Instituto de Investigaciones Agropecuarias (INIA): Chillán, Chile, 2000; 200p.

25. Agromet. Red Agrometeorológica de INIA. 2012. Available online: agromet.inia.cl (accessed on 10 December 2018).

26. Gutiérrez-Gamboa, G.; Garde-Cerdán, T.; Carrasco-Quiroz, M.; Pérez-Álvarez, E.P.; Martínez-Gil, A.M.; Del Alamo-Sanza, M.; Moreno-Simunovic, Y. Volatile composition of Carignan noir wines from ungrafted and grafted onto País (*Vitis vinifera* L.) grapevines from ten wine-growing sites in Maule Valley, Chile. *J. Sci. Food Agric.* **2018**, *98*, 4268–4278. [CrossRef]

27. Sadzawka, A.; Carrasco, R.; Grez, R.; Mora, M.; Flores, H.; Neaman, A. *Métodos de Análisis Recomendados para los Suelos de Chile*; Serie de Actas INIA, N° 34; Instituto de Investigaciones Agropecuarias: Santiago, Chile, 2006; pp. 164–165.

28. Gutiérrez-Gamboa, G.; Carrasco-Quiroz, M.; Martínez-Gil, A.M.; Pérez-Álvarez, E.P.; Garde-Cerdán, T.; Moreno-Simunovic, Y. Grape and wine amino acid composition from Carignan noir grapevines growing under rainfed conditions in the Maule Valley, Chile: Effects of location and rootstock. *Food Res. Int.* **2018**, *105*, 344–352. [CrossRef]

29. Köse, B. Phenology and rpening of *Vitis vinifera* L. and *Vitis labrusca* L. varieties in the maritime climate of samsun in Turkey's Black Sea Region. *S. Afr. J. Enol. Vitic.* **2014**, *35*, 90–102. [CrossRef]

30. Jones, G. Climate Change in the Western United States grape growing regions. *Acta Hortic.* **2005**, *689*, 41–59. [CrossRef]

31. Gutiérrez-Gamboa, G.; Verdugo-Vásquez, N.; Carrasco-Quiroz, M.; Garde-Cerdán, T.; Martínez-Gil, A.M.; Moreno-Simunovic, Y. Carignan phenolic composition in wines from ten sites of the Maule Valley (Chile): Location and rootstock implications. *Sci. Hortic.* **2018**, *234*, 63–73. [CrossRef]

32. Shellie, K.C. Water productivity, yield, and berry composition in sustained versus regulated deficit irrigation of Merlot grapevines. *Am. J. Enol. Vitic.* **2014**, *65*, 197–205. [CrossRef]

33. International Organitation Vineyard and Wine. *Compendium of International Methods of Wine and Must Analysis*; OIV: Paris, France, 2003.

34. Amerine, M.A. Quality control in the California wine industry. *J. Milk Food Technol.* **1972**, *35*, 373–377. [CrossRef]

35. Main, G.; Morris, J.; Striegler, K. Rootstock effects on Chardonel productivity, fruit, and wine composition. *Am. J. Enol. Vitic.* **2002**, *53*, 37–40.

36. Vasconcelos, M.; Castagnoli, S. Leaf canopy structure and vine performance. *Am. J. Enol. Vitic.* **2001**, *51*, 390–396.

37. Dokoozlian, N.K.; Kliewer, W.M. Influence of light on the grape berry growth and composition varies during fruit development. *J. Am. Soc. Hortic. Sci.* **1996**, *121*, 869–874. [CrossRef]

38. Malusà, E.; Laurenti, E.; Ghibaudi, E.; Rolle, L. Influence of organic and conventional management on yield and composition of grape CV. "Grignolino". *Acta Hortic.* **2014**, *640*, 135–141. [CrossRef]

39. Döring, J.; Frisch, M.; Tittmann, S.; Stoll, M.; Kauer, R. Growth, yield and fruit quality of grapevines under organic and biodynamic management. *PLoS ONE* **2015**, *10*, e0138445. [CrossRef]

40. Brunetto, G.; Ceretta, C.A.; de Melo, G.W.B.; Miotto, A.; Avelar Ferreira, P.A.; da Rosa Couto, R.; da Silva, L.O.S.; Garlet, L.P.; Somavilla, L.M.; Cancian, A.; et al. Grape yield and must composition of 'Cabernet Sauvignon' grapevines with organic compost and urea fertilization. *Rev. Cienc. Agrovet.* **2018**, *17*, 212–218. [CrossRef]

41. Sadras, V.; Moran, M.; Petrie, P. Resilience of grapevine yield in response to warming. *OENO One* **2017**, *51*, 381–386. [CrossRef]

42. William, L.E. Bud development and fruitfulness of grapevines. In *Raisin Production Manual*, 1st ed.; Christensen, L.P., Ed.; University of California, Agricultural and Nature Resources: Oakland, CA, USA, 2000; pp. 24–29, ISBN 1-879906-44-9.

43. Sánchez, L.A.; Dokoozlian, N.K. Bud microclimate and fruitfulness in *Vitis vinifera* L. *Am. J. Enol. Vitic.* **2005**, *56*, 319–329.

44. Bonada, M.; Sadras, V.O.; Fuentes, S. Effect of elevated temperature on the onset and rate of mesocarp cell death in berries of Shiraz and Chardonnay and its relationship with berry shrivel. *Aust. J. Grape Wine Res.* **2013**, *19*, 87–94. [CrossRef]

45. Tesic, D.; Keller, M.; Hutton, R.J. Influence of vineyard floor management practices on grapevine vegetative growth, yield, and fruit composition. *Am. J. Enol. Vitic.* **2007**, *58*, 1–11.

46. Villiers, F.S. The use of a Geographic Information System (GIS) in the selection of wines cultivars for specific areas by using temperature climatic models. In Proceedings of the XXII Congrès de la Vigne et du Vin, Buenos Aires, Argentina, 5 December 1997; Office International de la Vigne et du Vin: Paris, France, 1997.

47. Martin, K.R.; Rasmussen, K.K. Comparison of sensory qualities of geographically paired organic and conventional red wines from the southwestern US with differing total polyphenol concentrations: A randomized pilot study. *Food Nutr. Sci.* **2011**, *2*, 1150–1159. [CrossRef]

48. Koyoma, K.; Goto-Yamamoto, N. Bunch shading during different developmental stages affects the phenolic biosynthesis in berry skins of 'Cabernet Sauvignon' grapes. *J. Am. Soc. Hortic. Sci.* **2008**, *133*, 743–753. [CrossRef]

49. Cortell, J.M.; Kennedy, J.A. Effect of shading on accumulation of flavonoid compounds in (*Vitis vinifera* L.) Pinot noir fruit and extraction in a model system. *J. Agric. Food Chem.* **2006**, *54*, 8510–8520. [CrossRef]

50. Filippetti, I.; Allegro, G.; Valentini, G.; Pastore, C.; Colucci, E.; Intrieri, C. Influence of vigour on vine performance and berry composition of cv. Sangiovese (*Vitis vinifera* L.). *OENO One* **2003**, *47*, 21–33. [CrossRef]

51. Martínez-Gil, A.M.; Gutiérrez-Gamboa, G.; Garde-Cerdán, T.; Pérez-Álvarez, E.P.; Moreno-Simunovic, Y. Characterization of phenolic composition in Carignan noir grapes (*Vitis vinifera* L.) from six wine-growing sites in Maule Valley, Chile. *J. Sci. Food Agric.* **2017**, *98*, 274–282. [CrossRef]

52. Fließbach, A.; Eyhorn, F.; Mäder, P.; Rentsch, D.; Hany, R. DOK long-term farming systems trial: Microbial biomass, activity and diversity affect the decomposition of plant residues. In *Sustainable Management of Soil Organic Matter*; Rees, R.M., Ball, B.C., Campbell, C.D., Watson, C.A., Eds.; CABI: London, UK, 2001; pp. 363–369.

53. Morlat, R.; Chaussod, E. Long-term additions of organic amendments in a Loire Valley vineyard. I. Effects on properties of a calcareous sandy soil. *Am. J. Enol. Vitic.* **2008**, *59*, 353–363.

54. Steenwerth, K.; Belina, K.M. Cover crops enhance soil organic matter, carbon dynamics and microbiological function in a vineyard agroecosystem. *Appl. Soil Ecol.* **2008**, *40*, 359–369. [CrossRef]

55. Matus, F.J.; Rodriguez, J. A simple model for estimating the contribution of nitrogen mineralization to the nitrogen supply of crops from a stabilized pool of soil organic matter and recent organic input. *Plant Soil* **1994**, *162*, 259–271. [CrossRef]

56. Paustian, K.; Six, J.; Elliott, E.T.; Hunt, H.W. Management options for reducing CO_2 emissions from agricultural soils. *Biogeochemistry* **2000**, *48*, 147–163. [CrossRef]

57. Edwards, L.M. The effect of alternate freezing and thawing on aggregate stability and aggregate size distribution of some Prince Edward Island soils. *J. Soil Sci.* **1991**, *42*, 193–204. [CrossRef]

58. Cambardella, C.A.; Elliott, E.T. Carbon and nitrogen distribution in aggregates from cultivated and native grassland soils. *Soil Sci. Soc. Am. J.* **1993**, *57*, 1071–1076. [CrossRef]

59. Six, J.; Paustian, K.; Elliott, E.T.; Combrink, C. Soil Structure and Organic Matter. *Soil Sci. Soc. Am. J.* **2000**, *64*, 681–689. [CrossRef]

60. Muñoz, C.; Cruz, B.; Rojo, F.; Campos, J.; Casanova, M.; Doetterl, S.; Boeckx, P.; Zagal, E. Temperature sensitivity of carbon decomposition in soil aggregates along a climatic gradient. *J. Soil Sci. Plant Nutr.* **2016**, *16*, 461–476. [CrossRef]

61. Pocknee, S.; Sumner, M.E. Cation and nitrogen contents of organic matter determine its soil liming potential. *Soil Sci. Soc. Am. J.* **1997**, *61*, 86–92. [CrossRef]

62. Kätterer, T.; Reichstein, M.; Andrén, O.; Lomander, A. Temperature dependence of organic matter decomposition: A critical review using literature data analyzed with different models. *Biol. Fertil. Soils* **1998**, *27*, 258–262. [CrossRef]

63. Conant, R.T.; Ryan, M.G.; Ågren, G.I.; Birge, H.E.; Davidson, E.A.; Eliasson, P.E.; Bradford, M.A. Temperature and soil organic matter decomposition rates—Synthesis of current knowledge and a way forward. *Glob. Chang. Biol.* **2011**, *17*, 3392–3404. [CrossRef]

![agronomy logo] *agronomy*

MDPI

Article

PGR and Its Application Method Affect Number and Length of Runners Produced in 'Maehyang' and 'Sulhyang' Strawberries

Chen Liu [1], Ziwei Guo [1], Yoo Gyeong Park [2], Hao Wei [1] and Byoung Ryong Jeong [1,2,3,*]

[1] Department of Horticulture, Division of Applied Life Science (BK21 Plus Program), Graduate School of Gyeongsang National University, Jinju 52828, Korea; chenliu215@gmail.com (C.L.); guoziwei1230@gmail.com (Z.G.); oahiew@gmail.com (H.W.)

[2] Institute of Agriculture and Life Science, Gyeongsang National University, Jinju 52828, Korea; ygpark615@gmail.com

[3] Research Institute of Life Science, Gyeongsang National University, Jinju 52828, Korea

* Correspondence: brjeong@gmail.com; Tel.: +82-010-6751-5489

Received: 10 December 2018; Accepted: 25 January 2019; Published: 28 January 2019

Abstract: Vegetative propagation using runner plants is an important method to expand the cultivation area for the strawberry (*Fragaria* × *ananassa* Duch.). However, excessively long runners need an increased total amount of nutrients and energy to receive elongation from mother plants, which may lead to poor growth or reduced output. The use of plant growth regulators (PGRs) is an adoptable way to solve such problems. The objectives of this experiment were to study the effects of PGRs and their application methods on the growth and development of runners, runner plants, and mother plants, and also to find effective ways to control the number and length of runners without harmful side effects. Chlormequat chloride (CCC), 6-benzylaminopurine (BA), and ethephon (ETH) at a concentration of 100 mg·L^{-1} were applied via three different methods: injection into crowns, medium drench, and foliar spray. The results showed that BA injection into crowns was the most effective combination among all treatments, which prominently shortened the length of runners and increased the number of runners and leaves on a single plant. Furthermore, plants with BA solution injection tended to produce stronger runners with higher fresh and dry weights, without affecting the health states of mother plants. The ETH solution seemed to have toxic effects on plants, by leading many dead leaves and weak runners, and increased activities of antioxidant enzymes. Other than the injection method, the other two application methods of the CCC solution did not significantly affect the growth and development of both cultivars. Runner plants grown for 30 days were not affected by any treatments, and they were in similar conditions. Overall, BA injection into crowns is recommended for controlling the number and length of strawberry runners.

Keywords: chlormequat chloride; ethephon; *Fragaria* × *ananassa*; 6-benzylaminopurine; runner

---

1. Introduction

The strawberry (*Fragaria* × *ananassa*), a herbaceous perennial crop species from the Rosaceae family, is one of the most popular fruit crops with great economic values. As a berry, the strawberry is full of vitamins and minerals that are good for human health [1]. It is commercially grown in approximately 80 countries [2]. In 2013, global strawberry production exceeded 7.7 million tons [3]. Runners, or stolons, are stems that grow on the ground surface, with several nodes that are capable of generating adventitious roots and daughter plants because of their meristematic tissues during the growth and development stages [4]. Adventitious roots are available at the second, fourth, or sixth nodes, where the newly-formed plants can be used for propagation. However, in commercial production, runners and runner plants are generally removed by growers [5] because mother plants do not stop sending out runners and runners keep growing until some actions are taken, which may result in the deterioration of the mother plants' condition. Runners and runner plants need a large amount of productive energy and nutrition from mother plants, causing mother plants to have reduced outputs. It is also reported that asexual reproduction reduces fruit yields [6].

Although overgrowth of runners may influence the health of mother plants, as mentioned above, sometimes growers deliberately keep runners on mother plants to produce new generation plants. In the Republic of Korea, especially in the southern areas of the country, strawberry growers usually control environmental conditions to force strawberries to bloom as early as possible, usually by late October to early November, in the cultivation season. From November to the following February is the season to produce/harvest fruits. March to April is a suitable time for growing runners, while May and June are the best months to harvest runner plants to be used as new mother plants. Runner plants are often collected and sold or planted to expand the planting area. July to September is the time when the new generation plants grow. Such a strawberry production pattern forms a complete cycle in the Republic of Korea and is repeated year after year. Using runner plants for propagation is a method with easy operation, high propagation coefficient, and low cost. Cutting propagation by runner plants is an important way for strawberry reproduction because strawberry runner plants are easily produced and rooted [7], and the new plants retain their parents' good traits. To sum up, controlling the length and number of runners to limit harmful effects to their mother plants and offspring is important. Unfortunately, few studies have been carried out concerning this issue.

According to former research carried out by Kumar et al., potato runners can be controlled by phytohormones like indoleacetic acid (IAA) and gibberellins (GA) [8]. Plant growth regulator (PGR) is an artificial chemical phytohormone analogue that is essential for regulating plant growth and development in agriculture, and can be applied to control plant size, flowering, fruiting, and output [9]. It is also reported that PGRs, such as chlormequat chloride (commercially available under the trade name cycocel, CCC) and ethephon (ETH), can shorten certain parts of a plant [10]. Due to differences in properties and reactions of different plant tissues to PGRs, effects of PGRs may vary greatly. Even with the same PGR, different application methods may lead to different results. There are many methods of applying PGRs, such as spraying, drenching, and dipping. Therefore, it is worth trying because taking different approaches may yield unexpected benefits.

Discovered by Professor Tolbert at Michigan State University in the 1950s, CCC is the first plant growth regulator used on plants [11]. It is a synthetic PGR that is antagonistic to gibberellins (GA), while GAs are phytohormones that regulate plants' developmental processes, such as stem elongation, dormancy, and germination [12]. The CCC inhibits cell elongation without inhibiting cell division. Studies have shown that CCC can decrease the growth of stems, leaves, and runners of potato [13] and thicken the stem of mung beans [14] effectively by controlling vein growth and lodging. The CCC application results in dwarfed plants; thickened stalks [11]; darkened, greened, and thickened leaves; increased chlorophyll content; and a well-developed root system. Nowadays, CCC is widely used in agricultural production to slow down the stem growth, while enhancing flowering in many crops.

The ETH was discovered in 1965 and was first registered as a pesticide in the U.S. in 1973 [15]. It has low toxicity, and in the U.S. it is registered for use on ornamental plants as well as wheat, barley, apple, blackberry, cherry, grape, pineapple, cucumber, tomato, pepper, coffee, cotton, and tobacco [16]. The ETH is antagonistic to IAA, and is easily converted into ethylene in aqueous solutions of pH 5 or higher [17]. Ethylene interferes the growth processes of plants, and is a potent regulator of plant growth and ripeness. According to previous studies, exogenous ETH on turfgrass reduces the mowing frequency [18], which means it prevents or slows down the growth processes of turfgrass.

6-benzylaminopurine (BA) is a first generation man-made cytokinin that plays an important role in plant cell division, fruit growth acceleration, shoot formation, fruit setting, and yield increases [19]. Furthermore, BA increases plants' stress resistance [20] and therefore is widely used in horticulture and agriculture [21].

As discussed above, three PGRs have been widely researched, developed, and used commercially for the past half century to manipulate plant shape, form, and overall crop quality in agriculture and horticulture. The hypothesis of this study comes from the fact that some PGRs can control the length of plant organs, such as potato runners [13]. Especially, it is hypothesized that BA induces a large number of cells to divide, which would lead to more runners emerging and competing with each other, resulting in the decreased length of the runners. The objectives of this study were to realize the effects of PGRs and their application methods on the growth and development of strawberry runners, runner plants, and mother plants, and also to find out methods to control the number and length of runners without the harmful effects of PGRs on the plants.

2. Materials and Methods

2.1. Plant Materials and Culture Conditions

The strawberry cultivars used were 'Maehyang' and 'Sulhyang'. Plants were purchased from a strawberry farm (Sugok-myeon, Jinju, Gyeongsangnam-do, Republic of Korea) and maintained in the BVB Medium (Bas Van Buuren Substrate, EN-12580, De Lier, The Netherlands). The experiment was carried out in a glasshouse at Gyeongsang National University in the Republic of Korea. The culture environment had 23/17 °C day/night average temperatures, 70–80% relative humidity, and a natural photoperiod of 14 h or so. Plastic pots (Green-100, Danong Co., Namyangju, Republic of Korea) were used as growing containers, and plants were treated after confirming all plants produced at least one runner and the length of runner was approximately 5 cm. To count the number of new leaves and runners, the redundant leaves and runners were removed before the first treatment, leaving three leaves and one runner on each plant. Treatments were given weekly on Friday mornings for one month (1–31 May 2018).

When treatments were finished, 10 healthy and uniform runner plants in each treatment generated on the second node of runners were selected and stuck into the BVB Medium contained in 21-cell zigzag trays (21-Zigpot/21 cell tray, Daeseung, Jeonju, Republic of Korea). A fogging system (UH-303, JB Natural Co. Ltd., Gunpo, Republic of Korea) was used to promote induction of roots for about 9 days, and plants were checked to find out if there were any toxic effects from PGR treatments. The cultivation environment of runner plants had 32/21 °C day/night temperatures (average), 75–85% relative humidity, and a natural photoperiod of approximately 14.5 h. This stage lasted for another 30 days from June 1 to 30, 2018.

For maintenance, a greenhouse multipurpose nutrient solution (in $mg \cdot L^{-1}$ $Ca(NO_3)_2 \cdot 4H_2O$ 737.0, KNO_3 343.4, KH_2PO_4 163.2, K_2SO_4 43.5, $MgSO_4 \cdot H_2O$ 246.0, NH_4NO_3 80.0, Fe-EDTA 15.0, H_3BO_3 1.40, $NaMoO_4 \cdot 2H_2O$ 0.12, $MnSO_4 \cdot 4H_2O$ 2.10, and $ZnSO_4 \cdot 7H_2O$ 0.44 (electrical conductivity 0.8 $dS \cdot m^{-1}$)) was provided by drenching the growing medium daily.

2.2. Plant Growth Regulators Tested

CCC, BA, and ETH (MB-C4219, MB-B5812, and MB-E5360, respectively. MB Cell, Seoul, Republic of Korea) were used in this experiment. Unfortunately, there was little research done on the application of these PGRs and the proper concentration of these PGRs for strawberry. In accordance with agricultural production practices and guidelines for the use of PGRs, we found that growers usually treat cucumber, tomato, eggplant, and melon with CCC solution at 100–500 $mg \cdot L^{-1}$; rose and chrysanthemum with BA solution at 50–200 $mg \cdot L^{-1}$; and cucumber, melon, and watermelon at 100–500 $mg \cdot L^{-1}$. In order to prevent plant damage from highly concentrated PGR solutions and also to ensure the consistency of the experimental concentration, 100 $mg \cdot L^{-1}$ for all three PGRs was used in this experiment. The CCC and ETH were dissolved in distilled water, while BA was dissolved in a 1N NaOH solution. After testing, it was determined that application of a 5 mL of PGR for each treatment was the most appropriate for each plant, as 5 mL of the solution to a plant was the most appropriate, and 5 mL is sufficient to make the medium permeated and cover all the leaves of a plant without much loss.

2.3. Methods of Supplying PGR Solution

CCC, BA, and ETH at a concentration of 100 mg·L^{-1} were applied using three different mathods: injection into crowns, medium drench, and foliar spray.

2.3.1. Injection

To inject the PGR solution, crowns were pierced with an injection syringe (12 mL, Jung Rim Medical Industrial Co. Ltd., Seoul, Republic of Korea), and 5 mL solution was injected into the plant each time.

2.3.2. Drench

An injection syringe (25 mL, Jung Rim Medical Industrial Co. Ltd., Seoul, Republic of Korea) without a needle was used to drench the growing medium with 5 mL PGR solution.

2.3.3. Spray

For spraying, a sprayer was filled with the PGR solution at the total calculated volume according to the number of plants so that 5 mL is foliage-sprayed evenly per plant.

2.4. Measurements of Growth and Morphological Parameters

After 30 days of treatment, on 1 June 2018, growth parameters such as length of the longest runner, average length between the crown and the second node, runner diameter, number of runners per plant, fresh and dry weights of runners, and number of new leaves of the mother plants were measured. It is noteworthy that when the length of the longest runner was measured, most runners only had four nodes and very few of them had five nodes. To make the data more comparable, only length of four nodes was measured. After another 30 days, on 1 July 2018, growth parameters such as average length of shoot and root, crown diameter, fresh and dry weights of shoot and root, number of leaves, leaf length, leaf width and thickness, and chlorophyll level, of runner plants were measured.

Dry weights of shoot, root, and whole plant were measured after 72 h of drying in a drying oven (FO-450M, Jeio Technology Co. Ltd., Daejeon, Republic of Korea) at 70 °C. Diameters of the runner, crown, and leaf thickness were measured using a Vernier caliper (CD-20CPX, Mitutoyo Korea Co., Gunpo, Republic of Korea) at the widest points. The chlorophyll level was measured with a chlorophyll meter (SPAD-502, Konica Minolta Inc., Japan) on three healthy leaves in each plant to be averaged.

Samples for physiological analysis were taken from leaves of three randomly selected plants in each treatment among young and healthy leaves with uniform sizes and same conditions. Samples were fixed with liquid nitrogen as quickly as possible.

2.5. Measurements of Contents of Starch, Soluble Sugar, and Protein

2.5.1. Soluble Sugar and Starch

The contents of soluble sugar and starch were assayed according to the Anthrone colorimetric method [22]. For each treatment, a 0.2 g leaf sample was ground into the homogenate with distilled water and then transferred into a 15 mL centrifugal tube. The volume was adjusted to 6 mL and extracted in boiling water for 30 min. Then, the residue was filtered for starch extraction and the remaining solution was adjusted to 15 mL. A 0.1 mL of the extracted sample solution was added to 3 new 15 mL centrifugal tubes for each treatment, and 0.1 mL distilled water was used as the control. Then, 1.9 mL distilled water, 0.5 mL 2% Anthrone ethylacetate, and 5 mL 98% H_2SO_4 were added in that order into the tubes. All tubes were submerged in boiling water for 10 min. The absorbance was measured at 630 nm with a spectrophotometer (Uvikon 992, Kotron Instrumentals, Milano, Italy) after cooling the samples to room temperature.

The residue of sugars was used to assay the starch content. The residues from different treatments were added to tubes with 5 mL distilled water and boiled for 15 min to extract starch. Afterwards, 0.7 mL of 9.2 mol·L^{-1} perchloric acid was added to each tube, and tubes were placed in boiling water for additional 15 min. The volumes were adjusted to 15 mL after cooling down. The contents from the tubes were filtered, and the same volume of extracted samples, distilled water, Anthrone ethylacetate, and H$_2$SO$_4$ were added into new tubes to measure the absorbance at 485 nm. The soluble sugar and starch contents were calculated according to the prepared standard curves.

2.5.2. Protein

Total protein content was measured based on the reaction of Coomassie brilliant blue G-250 with proteins by measuring the absorbance at 595 nm with a spectrophotometer according to the method of Bradford [23]. The Na$_2$HPO$_4$ and NaH$_2$PO$_4$ were mixed in distilled water according to protocol for the phosphate buffer. Afterwards, 0.058 g EDTA-Na$_2$, 0.1 mL 0.05% Triton X solution, and 4.0 g 2% PVP were added to the phosphate buffer and the pH was adjusted to 7.0 to finish the working buffer. A 0.1 g of the leaf sample and 1.5 mL of working buffer were taken and ground into the homogenate in an ice box. This mixture was centrifuged at 13,000 rpm, 4 °C for 20 min with a centrifuge (5430 R, Eppendorf AG, Hamburg, Germany), and then the supernatant was transferred to new e-tubes. For measurement, 50 μL of the supernatant was mixed with 1,450 μL of Bradford's reagent, and was held still for 5–10 min. A standard curve was made by using Bovine serum albumin.

2.6. Measurements of Antioxidant Enzymes Activities

2.6.1. Superoxidase Dismutase (SOD)

According to the protocol of Beauchamp and Fridovich [24], the SOD activity was assayed by measuring the capacity to inhibit the photochemical reduction of nitroblue tetrazolium (NBT). The measurement was conducted with a 3 mL reaction mixture containing 50 mM phosphate buffer (pH 7.8), 14.5 mM methionine, 2.25 μM NBT, 60 μM riboflavin, 30 μM EDTA, and 0.1 mL of the enzyme extract. This reaction solution was incubated for 20 min under fluorescent lamps at an illuminance of 4000 lux. A tube containing the enzyme was kept in dark and served as the blank, while the control tube without enzyme extracts was kept in light. The absorbance was taken at 560 nm, and calculations were made by using an extinction coefficient of 100 mM^{-1}·cm^{-1}.

2.6.2. Peroxidase (POD)

According to the protocol described by Sadasivam and Manickam [25], a 0.2 M phosphate buffer (pH 6.0), 0.076 mL guaiacol solution, 0.1 mL enzyme extract, and 0.112 mL 30% hydrogen peroxide solution were prepared for the enzyme assay. An increase in the absorbance was recorded at 470 nm. The time was recorded in 30-s intervals until the decrease became constant. The extinction coefficient was 6.39 per micromole.

2.6.3. Catalase (CAT)

The total CAT activity was measured by the method of Aebi [26]. The assay system consisted of a 0.15 M phosphate buffer (pH 7.0), 0.31 mL 30% hydrogen peroxide solution, and 0.1 mL of the enzyme extract in the final volume of 3 mL. The decrease in the absorbance was recorded at 240 nm. The molar extinction coefficient of H$_2$O$_2$ at 240 nm was 0.004 μmol^{-1}·cm^{-1}.

2.7. Statistical Analysis

The data were analyzed with SAS (SAS 9.4, SAS Institute Inc., Cary, NC, USA). The experimental results were subjected to an analysis of variance (ANOVA) and Duncan's multiple range tests. OriginPro 9.0 (OriginLab Co., Northampton, MA, USA) was used for graphing.

3. Results

3.1. Effects of PGR and Application Method on Runners and Mother Plants

As shown in Figure 1 and Table 1, significant differences were observed between the two cultivars. Runners of strawberry 'Sulhyang' were much longer and stronger than those of 'Maehyang', and PGR affected growth and development of runners in terms of their length; diameter; and, especially, runner number. The application methods also had significant effects on growth and development of runners, such as number, length, diameter, and fresh weight. As for the effect of PGR in combination with application method, diverse effects on number, length, diameter, and dry weight of the runners were found. For example, plants treated with PGRs by injection method shortened length of runners.

For strawberry 'Maehyang', all three PGRs shortened runner length when injected. The treatment of BA injected into crowns induced the greatest number of runners and leaves per plant (Figure 1B). All treatments increased runner diameter, and it was most pronounced in the treatment of BA injected into crowns. The greatest fresh weight of runners was obtained in the treatments of ETH drench, BA injection, and BA drench among all treatments. The ETH solution at a concentration of 100 mg·L^{-1} may have had toxic or senescing effects on the mother plants, as its injection and drench caused many leaves to die (Figure 1C). Furthermore, ETH drench tended to prevent the formation of runners. For strawberry 'Sulhyang', all treatments except the BA injection and BA drench tended to prevent runner induction. The BA injection resulted in the greatest number of runners per plant (Figure 1B). The injection of either ETH or BA was effective in controlling length of runners. Fresh and dry weights of plants in all treatments were not significantly different from those in the control group. There was little difference in number of new leaves and runner diameter among treatments. Drenching medium with a 100 mg·L^{-1} ETH solution may have had toxic or senescing effects on the strawberry plants as it resulted in many dead leaves in this cultivar also (Figure 1C). In summary, injection of a 100 mg·L^{-1} BA solution into crowns of mother plants of both 'Maehyang' and 'Sulhyang' strawberry was effective in controlling number and length of the runners.

Table 1. Effect of plat growth regulators (PGR) and application method on growth and development of runners of 'Maehyang' and 'Sulhyang' strawberry measured at 30 days after treatment initiation.

Cultivar (C)	PGR (P)	Treatment Method (M)	Runner					
			Number	Length of the Longest Runner (cm)	Length of the 1st & 2nd Internode (cm)	Diameter (mm)	Fresh Weight (g)	Dry Weight (g)
'Maehyang'	CCC	Injection	1.7 ± 0.2 b [z]	74.1 ± 3.1 b	30.9 ± 2.0 b	2.40 ± 0.04 cd	5.17 ± 0.35 bc	0.27 ± 0.02 a
		Drench	1.5 ± 0.2 b	85.0 ± 5.8 ab	37.6 ± 4.0 ab	2.54 ± 0.10 bc	6.36 ± 0.52 a-c	0.32 ± 0.05 a
		Spray	2.0 ± 0.1 b	87.8 ± 2.4 a	38.5 ± 1.6 ab	2.42 ± 0.12 cd	5.56 ± 0.57 a-c	0.29 ± 0.07 a
	BA	Injection	3.5 ± 0.4 a	74.9 ± 2.7 b	32.1 ± 1.5 b	2.75 ± 0.24 a	6.80 ± 0.86 ab	0.39 ± 0.04 a
		Drench	1.7 ± 0.2 b	83.4 ± 2.1 ab	37.8 ± 1.3 ab	2.49 ± 0.06 bc	7.23 ± 0.83 a-c	0.35 ± 0.06 a
		Spray	1.8 ± 0.2 b	86.9 ± 4.0 a	40.0 ± 1.7 a	2.43 ± 0.08 cd	5.48 ± 0.30 a-c	0.25 ± 0.04 a
	ETH	Injection	1.7 ± 0.2 b	73.9 ± 2.1 b	31.3 ± 1.9 ab	2.41 ± 0.07 cd	5.29 ± 0.44 bc	0.26 ± 0.01 a
		Drench	1.3 ± 0.2 b	83.5 ± 6.5 ab	33.8 ± 3.1 ab	2.70 ± 0.08 ab	7.26 ± 0.80 a	0.32 ± 0.05 a
		Spray	1.5 ± 0.2 b	81.8 ± 3.6 ab	36.3 ± 1.3 ab	2.34 ± 0.10 cd	4.83 ± 0.57 c	0.27 ± 0.05 a
'Sulhyang'	CCC	Injection	2.0 ± 0.4 bc	72.8 ± 1.5 a-c	28.1 ± 1.8 ab	2.11 ± 0.13 c	6.10 ± 0.62 a-c	0.28 ± 0.02 bc
		Drench	2.2 ± 0.3 bc	81.3 ± 3.5 a	31.1 ± 2.8 ab	2.24 ± 0.09 a-c	3.27 ± 0.96 c	0.29 ± 0.02 a-c
		Spray	1.7 ± 0.3 bc	77.5 ± 4.8 ab	30.8 ± 2.2 ab	2.32 ± 0.09 a-c	7.06 ± 0.71 a	0.35 ± 0.03 ab
	BA	Injection	3.5 ± 0.2 a	64.2 ± 4.5 c	27.9 ± 1.3 ab	2.40 ± 0.10 a-c	3.45 ± 0.15 a-c	0.37 ± 0.03 a
		Drench	2.2 ± 0.2 bc	72.6 ± 2.2 a-c	34.0 ± 2.7 a	2.23 ± 0.09 a-c	5.41 ± 0.78 a-c	0.23 ± 0.04 c
		Spray	2.3 ± 0.4 b	79.8 ± 2.6 a	33.4 ± 1.2 a	2.18 ± 0.08 bc	4.53 ± 0.54 c	0.27 ± 0.04 bc
	ETH	Injection	1.3 ± 0.2 c	66.1 ± 3.6 c	25.5 ± 2.4 b	2.22 ± 0.13 a-c	4.94 ± 0.71 a-c	0.23 ± 0.03 c
		Drench	2.0 ± 0.3 bc	67.9 ± 3.5 bc	28.8 ± 2.0 ab	2.65 ± 0.13 a	6.69 ± 0.85 ab	0.27 ± 0.04 bc
		Spray	1.8 ± 0.3 bc	68.1 ± 2.9 bc	30.7 ± 1.3 ab	2.60 ± 0.09 ab	6.12 ± 0.60 a-c	0.31 ± 0.03 a-c
F-test [y]	C		*	***	***	**	NS	NS
	P		***	*	NS	*	NS	NS
	M		**	***	***	NS	**	NS
	C × P		NS	NS	NS	NS	NS	NS
	C × M		NS	NS	NS	NS	NS	NS
	P × M		***	*	NS	*	NS	*
	C × P × M		NS	NS	NS	NS	NS	NS

z Mean separation within columns for each cultivar by Duncan's multiple range test at $p < 0.05$. y NS, *, **, and ***: nonsignificant or significant at $p \leq 0.05$, 0.01, and 0.001, respectively.

Figure 1. The effects of (**A**) chlormequat chloride (CCC), (**B**) 6-benzylaminopurine (BA), and (**C**) ethephon (ETH) at a concentration of 100 mg·L^{-1} and the application method (injection, drench, and spray) on the number and length of runners of strawberries 'Maehyang' and 'Sulhyang' observed 30 days after treatment initiation in strawberries 'Maehyang' and 'Sulhyang': The letters C, B, and E stand for CCC, BA, and ETH, respectively; I, d, and s stand for injection into crowns, medium drench, and foliar spray, and Con. is the control.

3.2. The Effects of the PGR Solution and the Application Method on Endogenous Compounds

An analysis of the endogenous compounds showed that the starch, soluble sugar, and protein contents were differently affected by the treatments (Figure 2). The BA injection, BA foliar spray, and ETH injection resulted in lower contents of starch in strawberry 'Maehyang'. All CCC applications, regardless of the application method and BA injection, led to similarly lower starch contents in 'Sulhyang'. For 'Maehyang', CCC injection and all BA treatments resulted in the greatest soluble sugar contents, while BA drench, ETH drench, and ETH spray led to the greatest soluble sugar contents in 'Sulhyang'. The greatest protein contents were obtained in the CCC injection, BA injection, BA spray, and ETH injection for 'Maehyang' and in CCC injection, BA injection, and BA drench for 'Sulhyang'.

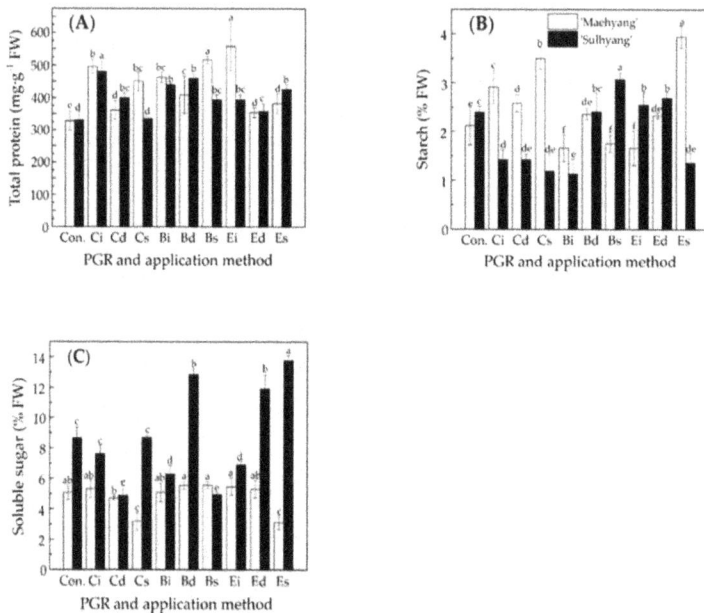

Figure 2. The effects of chlormequat chloride (CCC), 6-benzylaminopurine (BA), and ethephon (ETH) at a concentration of 100 mg·L^{-1} and their application method (injection, drench, and spray) on the content of (**A**) total protein, (**B**) starch, and (**C**) soluble sugars in leaves of strawberries 'Maehyang' and 'Sulhyang' sampled 30 days after treatment initiation. Vertical bars indicate the standard error ($n = 3$). Means accompanied by different letters are significantly different ($p < 0.05$) according to the Duncan's multiple range test: The letters C, B, and E stand for CCC, BA, and ETH, respectively; I, d, and s stand for injection into crowns, medium drench, and foliar spray, and Con. is the control.

3.3. The Effects of the PGR Solution and the Application Method on the Activities of Antioxidant Enzymes

All treatments resulted in increased SOD activity in 'Maehyang' compared to the control group. The CCC drench, ETH drench, and ETH spray resulted in markedly increased SOD activity compared to CCC injection, CCC spray, and BA injection. The SOD activity in 'Sulhyang' in the CCC spray, ETH drench, and ETH injection treatments was greater than that of the control group. The CCC drench and BA injection resulted in the lowest SOD activity. The CCC injection, CCC spray, and BA spray resulted in a considerably higher POD activity than other treatments in 'Maehyang'. Similarly, foliar spray of either CCC or BA resulted in much higher POD activity than other treatments, while all other treatments did not increase POD activity than the control in 'Sulhyang'. The CAT activity in 'Maehyang' in all treatments increased compared to the control, with the exception of BA injection and CCC drench, which had much higher CAT activity than the control. For 'Sulhyang', BA injection resulted in the lowest CAT activity, followed by CCC drench and CCC spray, and all other treatments led to high CAT activities (Figure 3).

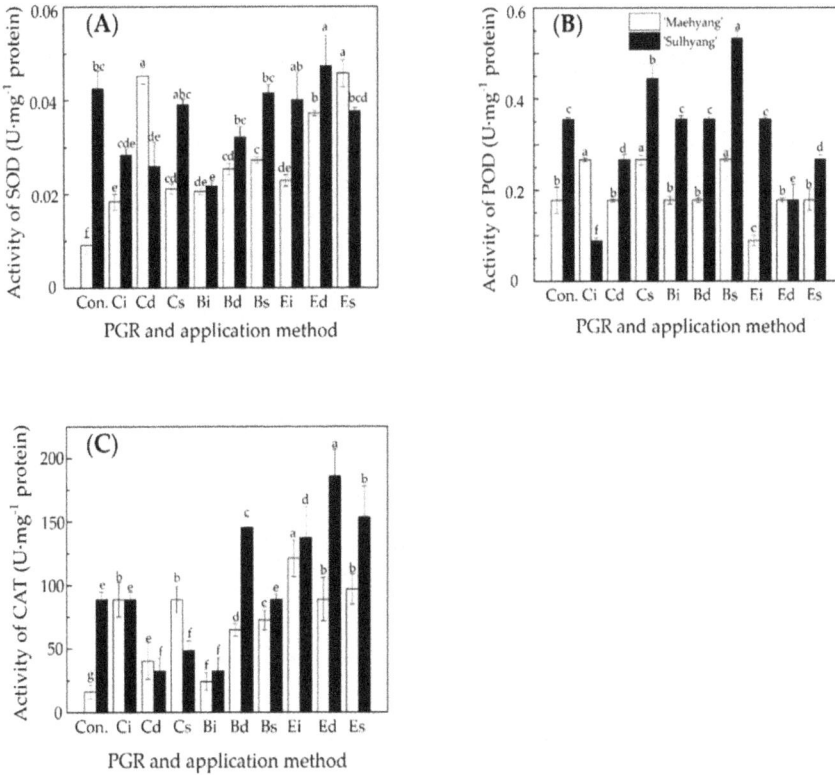

Figure 3. The effects of the chlormequat chloride (CCC), 6-benzylaminopurine (BA), and ethephon (ETH) at a concentration of 100 mg·L^{-1} and their application method (injection, drench, and spray) on the activities of (**A**) SOD (superoxidase dismutase), (**B**) POD (peroxidase), and (**C**) CAT (catalase) in the leaves of strawberry 'Maehyang' and 'Sulhyang' sampled 30 days after treatment initiation. Vertical bars indicate the standard error (*n* = 3). Means accompanied by different letters are significantly different (*p* < 0.05) according to the Duncan's multiple range test: The letters C, B, and E stand for CCC, BA, and ETH, respectively; I, d, and s stand for injection into crowns, medium drench, and foliar spray, and Con. is the control.

3.4. The Effects of the PGR Solution and the Application Method on the Runner Plants

With respect to the next generation, all runner plants with PGRs applied in different methods were observed to grow better than or at least as well as the control group for both cultivars. Across the two cultivars, there were significant differences that include the growth data of the shoot, root, and leaf, while within the same cultivar there was little difference between plants in different treatments (Figure 4). According to the results of the F-test, the PGRs and the application method exhibited little influence on the runner plants, which means that different PGR treatments in this study had no harmful effects on the next generation of strawberry plants. Data are shown in Tables 2 and 3.

Table 2. Effect of PGR and application method on growth and development of runner plants of strawberries 'Maehyang' and 'Sulhyang' grown for 30 days.

Cultivar (C)	PGR (P)	Application Method (M)	Shoot			Root	
			Length (cm)	Crown Diameter (mm)	Fresh Weight (g)	Length of Longest Root (cm)	Fresh Weight (g)
'Maehyang'	CCC	Injection	24.2 ± 1.0 a^z	7.20 ± 0.21 a	6.31 ± 0.32 a-c	18.2 ± 1.3	1.76 ± 0.14
		Drench	24.6 ± 1.2 a	7.13 ± 0.10 a	7.01 ± 0.24 ab	17.2 ± 0.7	1.84 ± 0.26
		Spray	24.4 ± 1.6 a	7.60 ± 0.37 a	7.28 ± 0.75 a	16.0 ± 1.2	1.75 ± 0.16
	BA	Injection	24.9 ± 1.0 a	7.36 ± 0.53 a	6.17 ± 0.39 a-c	18.3 ± 0.1	1.78 ± 0.20
		Drench	17.9 ± 0.6 d	7.02 ± 0.13 a	3.54 ± 0.15 d	16.9 ± 1.6	1.60 ± 0.14
		Spray	19.5 ± 2.2 cd	6.74 ± 0.24 b	5.01 ± 0.39 c	16.6 ± 0.3	1.86 ± 0.26
	ETH	Injection	24.0 ± 1.6 a	7.71 ± 0.36 a	6.83 ± 0.22 ab	17.7 ± 1.1	1.94 ± 0.16
		Drench	22.7 ± 1.8 ab	7.54 ± 0.41 a	7.08 ± 0.76 ab	19.1 ± 1.0	2.15 ± 0.27
		Spray	24.8 ± 1.0 a	7.19 ± 0.30 a	5.93 ± 0.34 a-c	18.0 ± 0.4	1.72 ± 0.21
'Sulhyang'	CCC	Injection	26.1 ± 1.3 ab	6.91 ± 0.13 ab	5.97 ± 0.92 ab	15.5 ± 2.0	1.70 ± 0.33
		Drench	26.1 ± 0.7 ab	6.61 ± 0.30 ab	5.46 ± 1.20 ab	17.8 ± 1.0	1.89 ± 0.30
		Spray	27.1 ± 1.7 a	7.20 ± 0.36 a	6.13 ± 0.70 ab	17.5 ± 0.6	2.11 ± 0.15
	BA	Injection	23.1 ± 0.1 b-d	6.82 ± 0.12 ab	4.21 ± 0.30 b	15.8 ± 0.8	1.93 ± 0.15
		Drench	27.3 ± 0.5 a	7.20 ± 0.39 a	7.14 ± 0.45 a	17.6 ± 0.1	1.94 ± 0.37
		Spray	22.1 ± 0.5 cd	6.85 ± 0.13 ab	3.88 ± 0.32 b	15.8 ± 0.4	1.96 ± 0.17
	ETH	Injection	24.4 ± 0.8 a-c	6.46 ± 0.24 ab	4.19 ± 1.79 b	16.2 ± 0.4	1.73 ± 0.30
		Drench	24.3 ± 0.4 a-c	5.43 ± 0.12 c	5.64 ± 0.47 ab	17.1 ± 0.6	1.91 ± 0.31
		Spray	21.1 ± 1.7 d	6.22 ± 0.33 b	4.24 ± 0.46 b	15.3 ± 0.7	1.59 ± 0.22
F-test^y	C		*	***	**	NS	NS
	P		***	NS	**	NS	NS
	M		NS	NS	NS	NS	NS
	C × P		*	***	*	NS	NS
	C × M		*	NS	NS	NS	NS
	P × M		NS	NS	*	NS	NS
	C × P × M		*	NS	*	NS	NS

z Mean separation within columns for each cultivar by Duncan's multiple range test at p < 0.05. y NS, *, **, and ***: nonsignificant or significant at p ≤ 0.05, 0.01, and 0.001, respectively.

Table 3. Effect of PGR and application method on growth and development of runner plants of strawberries 'Maehyang' and 'Sulhyang' grown for 30 days.

Cultivar (C)	PGR (P)	Method (M)	Number	The Largest Leaf				
				Length (cm)	Width (cm)	Thickness (mm)	Petiole Diameter (mm)	Chlorophyll (SPAD)
'Maehyang'	CCC	Injection	5.0 ± 0.0 a^z	7.2 ± 0.2 ab	4.8 ± 0.1 ab	0.55 ± 0.05	2.43 ± 0.08 ab	35.7 ± 2.1 b
		Drench	5.0 ± 0.0 a	7.5 ± 0.2 ab	5.4 ± 0.1 a	0.55 ± 0.04	2.44 ± 0.06 ab	38.8 ± 0.3 ab
		Spray	5.0 ± 0.0 a	7.5 ± 0.5 ab	5.2 ± 0.4 ab	0.53 ± 0.05	2.46 ± 0.12 ab	38.2 ± 1.2 ab
	BA	Injection	5.0 ± 0.0 a	7.5 ± 0.5 ab	5.5 ± 0.5 a	0.60 ± 0.04	2.44 ± 0.10 ab	39.5 ± 1.5 ab
		Drench	5.0 ± 0.0 a	5.2 ± 0.2 d	3.7 ± 0.1 c	0.51 ± 0.05	2.04 ± 0.14 c	38.5 ± 0.9 ab
		Spray	3.0 ± 0.0 c	5.8 ± 0.2 cd	4.2 ± 0.1 bc	0.51 ± 0.04	2.28 ± 0.05 bc	39.3 ± 0.6 ab
	ETH	Injection	4.0 ± 0.0 b	7.7 ± 0.7 a	5.3 ± 0.5 ab	0.54 ± 0.04	2.43 ± 0.11 ab	39.8 ± 0.8 a
		Drench	4.7 ± 0.3 a	6.9 ± 0.7 a–c	4.9 ± 0.1 ab	0.57 ± 0.04	2.66 ± 0.15 a	38.1 ± 1.5 ab
		Spray	4.7 ± 0.3 a	6.8 ± 0.7 a–c	4.9 ± 0.1 ab	0.53 ± 0.02	2.33 ± 0.07 a–c	39.6 ± 1.0 ab
'Sulhyang'	CCC	Injection	5.3 ± 0.3 ab	8.0 ± 0.2 a–d	5.6 ± 0.2 ab	0.34 ± 0.01	2.51 ± 0.18 a	32.6 ± 4.4 b
		Drench	5.0 ± 0.6 ab	7.7 ± 0.7 a–d	5.6 ± 0.6 ab	0.32 ± 0.01	2.22 ± 0.10 a–c	35.4 ± 1.7 ab
		Spray	5.7 ± 0.3 a	8.3 ± 0.5 a–c	5.8 ± 0.3 ab	0.33 ± 0.02	2.50 ± 0.06 a	37.2 ± 2.0 ab
	BA	Injection	5.3 ± 0.3 ab	6.9 ± 0.4 b–d	4.9 ± 0.2 b	0.35 ± 0.01	2.03 ± 0.01 c	36.2 ± 2.3 ab
		Drench	5.3 ± 0.3 ab	8.7 ± 0.6 a	6.2 ± 0.4 a	0.33 ± 0.02	2.49 ± 0.09 a	34.5 ± 0.7 ab
		Spray	5.3 ± 0.3 ab	6.7 ± 0.5 cd	4.9 ± 0.3 b	0.31 ± 0.03	1.96 ± 0.06 c	35.3 ± 1.5 ab
	ETH	Injection	5.7 ± 0.3 a	8.3 ± 0.4 ab	5.6 ± 0.3 ab	0.34 ± 0.02	2.37 ± 0.11 ab	40.2 ± 1.2 a
		Drench	5.0 ± 0.6 ab	7.8 ± 0.3 a–d	5.2 ± 0.2 ab	0.33 ± 0.02	2.38 ± 0.01 ab	37.9 ± 0.1 ab
		Spray	4.3 ± 0.3 b	6.6 ± 0.6 d	4.8 ± 0.5 b	0.31 ± 0.03	2.10 ± 0.16 bc	37.1 ± 2.4 ab
F-test^y	C		***	**	**	***	*	**
	P		*	**	*	NS	**	NS
	M		**	NS	NS	NS	NS	NS
	C × P		**	NS	NS	NS	NS	NS
	C × M		NS	*	NS	NS	NS	NS
	P × M		NS	NS	NS	NS	NS	NS
	C × P × M		*	*	**	NS	***	NS

z Mean separation within columns for each cultivar by Duncan's multiple range test at $p < 0.05$. y NS, *, **, and ***: nonsignificant or significant at $p \leq 0.05$, 0.01, and 0.001, respectively.

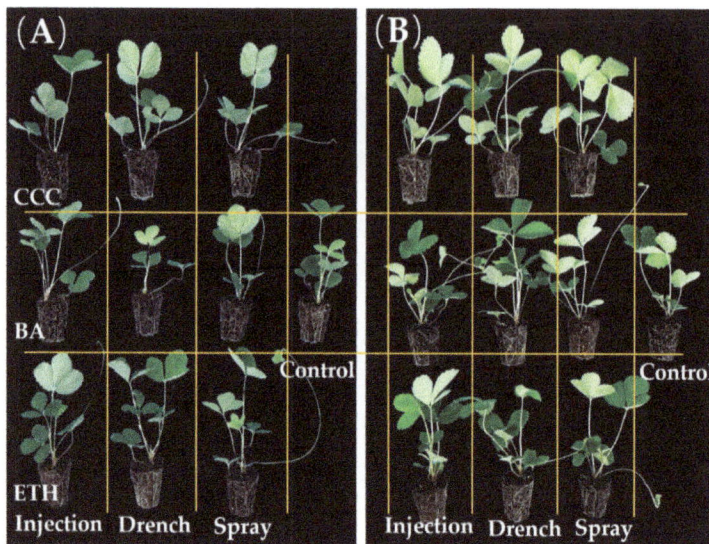

Figure 4. The effects of the chlormequat chloride (CCC), 6-benzylaminopurine (BA), and ethephon (ETH) at a concentration of 100 mg·L^{-1} and the application method (injection, drench, and spray) on the morphology of strawberries (**A**) 'Maehyang' and (**B**) 'Sulhyang' runner plants grown for 30 days.

4. Discussion and Conclusions

4.1. Discussion

Starch is the main form of storage carbohydrate in plants and is important in the carbon economy of many organs, tissues, and cell types of plants [27]. Soluble sugars play an important role in plant growth and developmental processes. They provide energy and mid-metabolites, and also act as signals, regulating the vital movements of plants [28]. The BA was reported to help in accumulating starch in *Lemna minor* [29] and ETH in increasing starch content in apple [30], which are in agreement with results of this study. Medium drenching or foliar spray of BA or ETH led to higher starch content in both cultivars of strawberry plants, while BA injection resulted in the lowest starch content for both cultivars. Root drenching or foliar spray of BA or ETH led to increased starch contents in both cultivars of strawberry plants, while BA injection decreased starch content to the lowest level for both cultivars. On the other hand, BA injection induced an increased level of soluble sugar content in both cultivars. Presumably, some functions of BA help in transforming starch into soluble sugars, and the intensity of this transformation is related to different plant tissues. It was observed that injection of any of three PGRs resulted in shortened runner length for both cultivars, while drenching or spraying PGRs did not. This indicates that PGRs have different mobile characteristics and induced effects depending on the location on the strawberry plants where they are applied, just as hypothesized. The actual mechanisms behind this need to be investigated further.

It is reported that CCC is highly mobile in both xylem and phloem tissues, and is rapidly absorbed and translocated [31]. However, CCC shows different mobility characteristics depending on the species. It was observed that CCC had high mobility in wheat but slow uptake/movement in barley, making CCC more effective in wheat than in barley [32]. In this study, strawberry plants with CCC treatments displayed a similar number of new leaves and runners, length of the first node, percentage of plants with new runners, and runner dry weights, while no significant differences were observed among the plants with respect to the application method. Therefore, it can be concluded that CCC is highly mobile in strawberries too. However, CCC injection into crowns resulted in a more pronounced shortening of the runner length compared to foliar spray or medium drench. This is probably because CCC injection works directly on the crown where runners are generated, while foliar spray and medium drench result in reduced CCC dosage during the transportation process, making CCC relatively less effective in shortening the runner length.

Truernit et al. (2006) found in *Arabidopsis thaliana* that cytokinin regulates the expression of invertase and transports hexose [33]. As a kind of cytokinin, BA has the effect of inhibiting chlorophyll, nucleic acid, and protein decomposition in the leaves, and various functions such as transporting amino acids, auxin, and mineral salts to plant parts exposed to it [34]. Another important characteristic of BA is its poor mobility in plants; furthermore, its physiological effects are limited to the treatment site and its vicinity. That is the reason why BA injection into crowns induced many more new runners and leaves but medium drench and foliar spray with BA did not. Due to BA helping in accumulating nutrients, runners of plants injected with BA were much stronger. The competition for nutrients among these strong runners made them have a shorter average length compared to runners under other treatments. As a benefit, growers have more options to get healthy and strong runners, while also saving the nutrients and energy to support the mother plant.

The ETH is quite different from the other two PGRs mentioned above, as it eventually decomposes into ethylene, so temperature (high temperatures accelerate ETH movement inside the plant), pH, period of usage, and the plant growth stage can easily influence the effect of ETH [35]. In plant production settings, ETH is applied by spraying, dipping, smearing, or air fumigation, among which spraying is the most commonly used. The ETH can be absorbed by leaves, stems, fruits, and other organs but is mainly taken up through leaf surface absorption. When ETH enters the vascular bundle, it is transported to other tissues and organs as the organic matter moves, so ETH has some mobility within the plant [36]. The ETH that enters the cell is broken down gradually to release ethylene, and then produces its effect on the plant. The ETH is similar to ethylene in that it enhances the synthesis of RNA in cells and promotes the synthesis of proteins. It has also been shown by studies that plants treated with ETH had high protein contents (Figure 2A).

Both crown injection and medium drench of ETH induced dead leaves, albeit to different degrees, and inhibited formation of new leaves and runner in strawberries. Foliar ETH spray did not induce any leaf deaths. The other two PGRs did not have the same leaf-killing and development-inhibiting effects. These results indicate that ETH has differing effects that depend on the application site, or that different plant tissues or organs have different tolerances to ETH, as they do to certain natural phytohormones. Another possible explanation for the differing effects of ETH by application method is the concentration of the solution. In this study, a concentration at 100 mg·L^{-1} was used. The higher the applied concentration, the more likely the occurrence of phytotoxicity. Phytohormones are characterized by low doses but high efficiencies. Some of them, such as IAA, have effects that depend on the concentration, where they promote plant growth at a low concentration but inhibit growth at a high concentration. As an analogue of plant hormone, ETH may have similar characteristics. To clarify the reason and mechanism by which dead leaves were induced by crown injection and medium drench of ETH in this experiment, further research is needed.

Stressful environments can cause the accumulation of reactive oxygen species (ROS) or free radicals in plants. Harsh environments, normal oxygen metabolism, certain chemical reactions, or toxic agents in the environment could force plants to produce such substances that continuously threaten the cells and tissues. The ROS and free radicals are able to disrupt the metabolic activity and cell structure. When this occurs, additional free radicals are produced in a chain reaction that leads to more extensive damage to plants, particularly the oxidation of DNA, proteins, and membrane lipids. Fortunately, plants can defend themselves against such damages via synthesizing antioxidant enzymes such as SOD, POD, and CAT to eliminate stresses [37]. It is true that the use of PGRs could push plants into stressed states. One example is the dead leaves and a low number of newly-grown runners in plants treated with ETH as discussed above. The experimental results also confirmed that the treatments did cause some biological stresses on the strawberry plants, because the activities of antioxidant enzymes in most of the treated plants were higher than that in the control group. Generally, ETH treatments resulted in higher activities of antioxidant enzymes, and BA treatments, especially by injection, resulted in lower activities of antioxidant enzymes in strawberry. To some extent, BA injection had little toxic and side effects on strawberry mother plants.

Although the PGRs selected were reported to have low toxicity, according to a research in mice, ETH could be harmful for the kidney and liver even in small doses [38]. The CCC was also reported to be toxic on the fertility of mammals such as pigs and mice [39]. The BA toxicity is seldom reported, and thus can be considered as no concern for human and animal safety. For plant safety, improper application or excessively high concentrations of CCC result in severe marginal leaf chlorosis or chlorotic spotting [40].

Typical application methods for PGRs are foliar spray or medium drench, but substrate spray, bulb spray, seed soak, cutting, and liner dips are also used. Each method has advantages and disadvantages, so appropriate methods should be chosen for a particular situation [41]. The injection method used in the experiment was actually similar with cutting and liner dips in that it damages the surface of plant tissues and lets the PGRs into the plant body. It is not easy to inject PGR solutions into strawberry crowns because strawberry do not have blood vessels like human or animals. If we obey the definition of 'injection' strictly, namely, it involves pricking the surface tissues of plants with an injector and then sending the solution inside, as tested, one person can finish one or two plants per minute, and this was the method tried prior to this experiment.

In the second turn of the experiment, a new method for raising efficiency was found and used. An injector was used to make small pores on the crowns, pushed in to let its contents out, without the need for pricking the entire needle inside. This allowed solutions to enter the plant through small pores, and the effectiveness was not compromised at all, compared with the first method in prior experiment discussed above. This allowed for three-four additional plants to be treated per minute. This means that, for example, if a grower wished to control the number and length of 5000 strawberry plants, it would take 1000 man/min (16.7 man/h) under ideal conditions, making this improved injection method practically applicable to real production environments.

4.2. Conclusions

The three PGRs had different effects on strawberry runners and mother plants, and the results are variable as affected by their application methods. The most successful combination of PGR and application method in this study was BA injection into crowns, because it achieved expected experimental goal of shortening the average length of runners without harming the mother plants and daughter plants, while increasing the number of new leaves and runners, and the diameter of runners. It is important to note that BA has low toxicity, thus posing little to no risks to human health. Because the aforementioned injection method is easy and efficient, it is possible to lower the cost and time in production setting. Thus, BA injection into crowns is recommended for growers in need of controlling the number and length of runners. As for the other two PGRs, ETH treatment caused many dead leaves and weak runners in mother plants, and it had little effect in controlling runners; CCC solution with injection could shorten runner length, but the effect was less dramatic than BA injection. The other two methods were not observed having any special effects on either cultivar. Runner plants (next generation) grown for 30 days were not affected by any treatments, and they were in similar conditions. The effects of the three PGRs in different concentrations have not been tested yet. A 100 mg·L^{-1} is not necessarily the optimal concentration for the PGRs, and the PGR solutions with different concentrations should also lead to different results. The amount of PGR residue in the fruits has not been tested either, and may be the focus of future experiments.

Author Contributions: Conceptualization, B.R.J.; Methodology, B.R.J. and C.L.; Formal Analysis, C.L., Z.G., and H.W.; Resources, B.R.J.; Data Curation, Y.G.P.; Writing–Original Draft Preparation, C.L.; Writing–Review and Editing, B.R.J. and Y.G.P.; Project Administration, B.R.J.; Funding Acquisition, B.R.J., C.L., Z.G., Y.G.P., and H.W.

Funding: This research was funded by the Agrobio-Industry Technology Development Program; Ministry of Food, Agriculture, Forestry, and Fisheries; Republic of Korea (Project No. 315004-5). C.L., Z.G., and H.W. were supported by a scholarship from the BK21 Plus Program, Ministry of Education, Republic of Korea.

Acknowledgments: This research was supported by the Agrobio-Industry Technology Development Program; Ministry of Food, Agriculture, Forestry, and Fisheries; Republic of Korea (Project No. 315004-5). Chen Liu, Ziwei Guo, and Hao Wei were supported by a scholarship from the BK21 Plus Program, Ministry of Education, Republic of Korea.

Conflicts of Interest: The authors declare no conflict of interest.

References

1. Giampieri, F.; Tulipani, S.; Alvarez-Suarez, J.M.; Quiles, J.L.; Mezzetti, B.; Battino, M. The strawberry: Composition, nutritional quality, and impact on human health. *Nutrition* **2012**, *28*, 9–19. [CrossRef] [PubMed]
2. Torrico, A.; Salazar, S.; Kirschbaum, D.; Conci, V. Yield losses of asymptomatic strawberry plants infected with strawberry mild yellow edge virus. *Eur. J. Plant Pathol.* **2018**, *150*, 983–990. [CrossRef]
3. Simpson, D. The economic importance of strawberry crops. In *The Genomes of Rosaceous Berries and Their Wild Relatives*; Springer: Cham, Switzerland, 2018; pp. 1–7.
4. Xu, Q.; Fan, N.; Zhuang, L.; Yu, J.; Huang, B. Enhanced stolon growth and metabolic adjustment in creeping bentgrass with elevated CO_2 concentration. *Environ. Exp. Bot.* **2018**, *155*, 87–97. [CrossRef]
5. Park, S.W.; Kwack, Y.; Chun, C. Growth and propagation rate of strawberry transplants produced in a plant factory with artificial lighting as affected by separation time from stock plants. *Hortic. Environ. Biotechnol.* **2018**, *59*, 199–204. [CrossRef]
6. Barrett, S.C. Influences of clonality on plant sexual reproduction. *Proc. Natl. Acad. Sci. USA* **2015**, *112*, 8859–8866. [CrossRef] [PubMed]
7. Dolgun, O. Field performance of organically propagated and grown strawberry plugs and fresh plants. *J. Sci. Food Agric.* **2007**, *87*, 1364–1367. [CrossRef]
8. Kumar, D.; Wareing, P. Factors controlling stolon development in the potato plant. *New Phytol.* **1972**, *71*, 639–648. [CrossRef]
9. Nickell, L.G. Plant growth regulators. *Chem. Eng. News* **1978**, *56*, 18–34. [CrossRef]
10. Rajala, A.; Peltonen-Sainio, P. Plant growth regulator effects on spring cereal root and shoot growth. *Agron. J.* **2001**, *93*, 936–943. [CrossRef]
11. Lindstrom, R.; Tolbert, N. (2-Chloroethyl) trimethylammonium chloride and related compounds as plant growth substances. IV. Effect on chrysanthemums and poinsettias. *Q. Bull. Mich. State Univ. Agric. Exp. Stn.* **1960**, *42*, 917–928.
12. Hedden, P.; Sponsel, V. A century of gibberellin research. *J. Plant Growth Regul.* **2015**, *34*, 740–760. [CrossRef] [PubMed]
13. Sharma, N.; Kaur, N.; Gupta, A.K. Effect of chlorocholine chloride sprays on the carbohydrate composition and activities of sucrose metabolising enzymes in potato (*Solanum tuberosum* L.). *Plant Growth Regul.* **1998**, *26*, 97–103. [CrossRef]
14. Farooq, U.; Bano, A. Effect of abscisic acid and chlorocholine chloride on nodulation and biochemical content of *Vigna radiata* L. under water stress. *Pak. J. Bot.* **2006**, *38*, 1511–1518.
15. Strydhorst, S.; Hall, L.; Perrott, L. Plant growth regulators: What agronomists need to know. *Am. Soc. Agron.* **2018**, *51*, 22–26. [CrossRef]
16. Gianessi, L.P.; Marcelli, M.B. *Pesticide Use in US Crop Production: 1997*; National Center for Food and Agricultural Policy: Washington, DC, USA, 2000; p. 32.
17. Wertheim, S.; Webster, A. Manipulation of growth and development by plant bioregulators. In *Fundamentals of Temperate Zone Tree Fruit Production*; Backhuys Publishers: Leiden, The Netherlands, 2005; pp. 267–294.
18. Fishel, F.M. *Plant Growth Regulators*; Document PI-139; Pesticide Information Office, Florida Cooperative Extension Service, Institute of Food and Agricultural Sciences, University of Florida: Gainesville, FL, USA, 2006; p. 3.
19. Zhu, X.; Zeng, Y.; Zhang, Z.; Yang, Y.; Zhai, Y.; Wang, H.; Liu, L.; Hu, J.; Li, L. A new composite of graphene and molecularly imprinted polymer based on ionic liquids as functional monomer and cross-linker for electrochemical sensing 6-benzylaminopurine. *Biosens. Bioelectron.* **2018**, *108*, 38–45. [CrossRef]

20. Chen, B.; Yang, H. 6-Benzylaminopurine alleviates chilling injury of postharvest cucumber fruit through modulating antioxidant system and energy status. *J. Sci. Food Agric.* **2013**, *93*, 1915–1921. [CrossRef] [PubMed]

21. Zheng, M.; He, J.; Wang, Y.; Wang, C.; Ma, S.; Sun, X. Colorimetric recognition of 6-benzylaminopurine in environmental samples by using thioglycolic acid functionalized silver nanoparticles. *Spectrochim. Acta Part A Mol. Biomol. Spectrosc.* **2018**, *192*, 27–33. [CrossRef]

22. Dubois, M.; Gilles, K.; Hamilton, J.K.; Rebers, P.A.; Smith, F. A colorimetric method for the determination of sugars. *Nature* **1951**, *168*, 167. [CrossRef]

23. Bradford, M.M. A rapid and sensitive method for the quantitation of microgram quantities of protein utilizing the principle of protein-dye binding. *Anal. Biochem.* **1976**, *72*, 248–254. [CrossRef]

24. Beauchamp, C.; Fridovich, I. Superoxide dismutase: Improved assays and an assay applicable to acrylamide gels. *Anal. Biochem.* **1971**, *44*, 276–287. [CrossRef]

25. Sadasivam, S. *Biochemical Methods*; New Age International: Seborga, Italy, 1996; pp. 108–109.

26. Ahmad, P.; Jhon, R. Effect of salt stress on growth and biochemical parameters of *Pisum sativum* L. *Arch. Agron. Soil Sci.* **2005**, *51*, 665–672. [CrossRef]

27. Smith, A.M.; Zeeman, S.C. Quantification of starch in plant tissues. *Nat. Protoc.* **2006**, *1*, 1342. [CrossRef] [PubMed]

28. Wang, J.; Tang, Z. The regulation of soluble sugars in the growth and development of plants. *Bot. Res.* **2014**, *3*, 71–76.

29. Jong, J.G.T.D.; Veldstra, H. Investigations on cytokinins. I. Effect of 6-benzylaminopurine on growth and starch content of *Lemna minor*. *Physiol. Plant.* **1971**, *24*, 235–238. [CrossRef]

30. Wang, Z.; Dilley, D.R. Aminoethoxyvinylglycine, combined with ethephon, can enhance red color development without over-ripening apples. *HortScience* **2001**, *36*, 328–331.

31. Emam, Y.; Karimi, H. Influence of chlormequat chloride on five winter barley cultivars. *Iran Agric. Res.* **1996**, *15*, 101–114.

32. Lord, K.; Wheeler, A. Uptake and movement of ^{14}C-chlormequat chloride applied to leaves of barley and wheat. *J. Exp. Bot.* **1981**, *32*, 599–603. [CrossRef]

33. Truernit, E.; Siemering, K.R.; Hodge, S.; Grbic, V.; Haseloff, J. A map of KNAT gene expression in the *Arabidopsis* root. *Plant Mol. Biol.* **2006**, *60*, 1–20. [CrossRef] [PubMed]

34. Wingler, A.; Schaewen, A.; Leegood, R.C.; Lea, P.J.; Quick, W.P. Regulation of leaf senescence by cytokinin, sugars, and light: Effects on NADH-dependent hydroxypyruvate reductase. *Plant Physiol.* **1998**, *116*, 329–335. [CrossRef]

35. Saltveit, M.E. Postharvest Biology and Handling of Tomateos. In *Tomatoes*; Wageningen University and Research: Wageningen, The Netherland, 2018; pp. 309–323.

36. Foster, K.R.; Reid, D.M.; Pharis, R.P. Ethylene biosynthesis and ethephon metabolism and transport in barley. *Crop Sci.* **1992**, *32*, 1345–1352. [CrossRef]

37. Jo, E.H.; Soundararajan, P.; Park, Y.G.; Jeong, B.R. Effect of silicon on growth and tolerance of *Torenia fournieri* in vitro to NaCl stress. *Flower Res. J.* **2018**, *26*, 68–76. [CrossRef]

38. Yazar, S.; Baydan, E. The subchronic toxic effects of plant growth promoters in mice. *Ankara Univ. Vet. Fak. Derg.* **2008**, *55*, 17–21.

39. Sørensen, M.T.; Danielsen, V. Effects of the plant growth regulator, chlormequat, on mammalian fertility. *Int. J. Androl.* **2006**, *29*, 129–133. [CrossRef] [PubMed]

40. Kaczperski, M.P.; Armitage, A.M.; Lewis, P.M. Accelerating growth of plug-grown pansies with carbon dioxide and light. *HortScience* **1994**, *29*, 442.

41. Boldt, J.L. Whole Plant Response of Chrysanthemum to Paclobutrazol, Chlormequat Chloride, and (s)-Abscisic Acid as a Function of Exposure Time Using a Split-Root System. Ph.D. Dissertation, University of Florida, Gainesville, FL, USA, 2008; pp. 15–17.

agronomy

MDPI

Article

Bird Management in Blueberries and Grapes

Catherine A. Lindell [1,2,*], Melissa B. Hannay [1] and Benjamin C. Hawes [2]

[1] Department of Integrative Biology, Michigan State University, 288 Farm Ln., East Lansing, MI 48824, USA; bradymel@msu.edu
[2] Center for Global Change and Earth Observations, Michigan State University, 1405 S. Harrison Rd., East Lansing, MI 48823, USA; bhawes1@kent.edu
* Correspondence: lindellc@msu.edu; Tel.: +1-517-884-1241

Received: 8 November 2018; Accepted: 1 December 2018; Published: 7 December 2018

Abstract: Bird damage to fruit is a long-standing challenge for growers that imposes significant costs because of yield losses and grower efforts to manage birds. We measured bird damage in 'Bluecrop' blueberry fields and Pinot noir vineyards in 2012–2014 in Michigan to investigate how year, grower, and forest cover influenced the proportions of bird damage. We tested whether inflatable tubemen (2013–2014) and a methyl anthranilate spray (2015) reduced bird damage in blueberries, and tested the deterrent effect of inflatable tubemen in grapes (2014). Years when crop yield was lower tended to have a higher damage percentage; for blueberries, bird damage was highest in 2012, and in grapes, damage was highest in 2012 and 2014. Neither blueberry fields nor vineyards with inflatable tubemen showed significantly reduced bird damage, although the blueberry fields showed a non-significant trend toward lower damage in the tubemen blocks. Blueberry field halves treated with the methyl anthranilate spray had equivalent bird damage to untreated halves. Our results correspond to previous work showing that percent bird damage varies by year, which was likely because bird consumption of fruit is relatively constant over time, while fruit yield varies. Fruit growers should expect a higher proportion of bird damage in low-fruit contexts, such as low-yield years, and prepare to invest more in bird management at those times. Investigating patterns of bird damage and testing deterrent strategies remain challenges. Bird activity is spatially and temporally variable, and birds' mobility necessitates tests at large scales.

Keywords: fruit; Michigan; inflatable tubemen; methyl anthranilate; bird deterrent

1. Introduction

Cultivated blueberry (*Vaccinium corymbosum* L.) production in the United States (U.S.) more than doubled between 2005–2016 to nearly 272 million kilograms (600 million pounds), which was valued at over 700 million dollars [1,2]. U.S. grape (*Vitis vinifera* L.) production stayed steady between 2005–2016, at somewhat over six million metric tons (seven and a half million short tons), although the value nearly tripled to a price of $1500/short ton [1,2]. Fruit-eating birds pose consistent challenges for fruit growers (e.g., Tracey et al. [3] and Lindell et al. [4]). Growers from Michigan, New York, Oregon, California, and Washington estimated that bird damage to blueberries in 2011 was between 3.8–18.2%, and bird damage to wine grapes was between 2.9–9.2% [5]. Using these estimates with state price and production data, we calculated that, for example, bird damage costs Michigan blueberry growers over $14 million annually, and California wine grape growers over $49 million [5]. Given the size of the blueberry and wine grape industries, the documented health benefits of fruit and vegetable consumption [6,7], and the yield loss to pest birds, improving the understanding and management of pest birds will have economic benefits for both growers and consumers and increase the health of society.

In previous work on tree fruits, we found that the proportion of bird damage was higher in low-fruit yield contexts. For example, the proportion of bird damage to Michigan sweet cherries

(*Prunus avium* L.) was much higher in a low-yield year than in higher-yield years [4]. We also found some evidence of spatial influences; bird damage to sweet cherries was lower in blocks surrounded by other sweet cherries than in blocks surrounded by non-sweet cherry land-cover types. Local populations of fruit-eating birds are likely relatively constant over a year-to-year time frame in the study region, and so the amount of food that they consume is an absolute amount. When there is a small absolute amount of fruit available (low yield years) and/or this is the only field in the local area that offers fruit (no surrounding fruit fields), the fixed absolute amount that birds eat will appear as a proportionally higher amount of the available fruit.

We also found greater bird damage in both sweet and tart cherry (*Prunus cerasus* L.) blocks that were in landscapes with low forest cover (less than 50% forest cover, compared to blocks with greater than 50% forest cover), potentially because of the close proximity of important resources in low forest cover landscapes, i.e., fruit in the blocks and nesting habitat, and cover from predators in the forest [4]. This pattern has been documented previously: Great-tailed grackles (*Quiscalus mexicanus*) caused higher bird damage in grapefruit groves close to sugarcane fields in Florida, presumably because the sugarcane fields provide roost sites for the grackles [8]. Similarly, blackbirds cause more damage in sunflower fields near cattail marshes that provided roosting habitat compared to fields farther from marshes [9].

Grower management may be another influence on bird damage levels, although not necessarily in an intuitive manner. Our recent work (Elser et al. in prep) indicated that higher bird damage levels are sometimes positively associated with bird management. We believe this pattern results because growers are more likely to employ management when a block is particularly susceptible to bird damage. Thus, when a grower manages for pest birds, he/she may simply be able to reduce bird damage down to regional averages. Assessing the efficacy of bird management techniques is inherently difficult, because birds are highly mobile. Thus, control and treatment sites should be large, and have baseline levels of damage that are similar. In the present work, we relied on bird damage estimates of blocks in previous years, or grower information, to match control and treatment blocks.

A relatively new bird management technique is inflatable tubemen that engage in haphazard movements and are driven by a fan, which may make them more effective at reducing bird activity than traditional stationary scarecrows or other visual deterrents. Bird-deterrent sprays are an appealing possibility to some growers, because the application technique is familiar. The only chemical currently registered for use on fruit is methyl anthranilate. Methyl anthranilate is a compound that occurs in grapes and strawberries and is added to foods to provide a fruit flavor and odor. Some anthranilate derivatives are avoided by birds in laboratory situations [10]. Methyl anthranilate is the active ingredient in a number of bird-deterrent sprays, although strong evidence is lacking that it reduces bird activity in crops (e.g., Avery et al. [11] and Dieter et al. [12]).

Our specific objectives were to (1) investigate the influence of year, grower, and forest cover on proportions of bird damage in blueberries and grapes, (2) test the efficacy of inflatable tubemen in reducing bird damage in blueberries and grapes, (3) determine whether bird damage was reduced for blueberries treated with Avian Control®, Stone Soap Co. Inc., Sylvan Lake, MI, USA, a spray with methyl anthranilate as the active ingredient, compared to untreated blueberries, and (4) make recommendations about managing pest birds in fruit.

2. Materials and Methods

2.1. Study Regions

Our blueberry study region is in western Michigan at approximately 42.3° latitude, −86.1° longitude. The average annual temperature for this region is 10.2 °C, and the average annual precipitation is 947 mm. Our grape study region is in the northern lower peninsula of Michigan at approximately 44.9° latitude, −85.7° longitude. Average annual temperature for this region is 7.4 °C, and average annual precipitation is 841 mm. Detailed information on the counties, varieties of fruit, and years the studies were conducted is in Table 1.

Table 1. Summary of years, locations of studies in Michigan, and varieties of fruit.

Crop	Study	Counties	Varieties of Fruit	Years
Blueberries	Influences on bird damage	Van Buren, Allegan, Berrien	Bluecrop	2012–2014
	Inflatable tubemen	Van Buren, Eaton, Ottawa, Berrien	Bluecrop, Jersey	2013–2014
	Methyl anthranilate	Van Buren	Bluecrop	2015
Grapes	Influences on bird damage	Leelanau, Grand Traverse	Pinot noir	2012–2014
	Inflatable tubemen	Leelanau, Grand Traverse	Pinot noir	2014

2.2. Block Selection

We defined a block as a contiguous area of one crop, with edges delimited by other land-cover types at least five meters wide. We approached fruit growers in each region to gain access to commercial orchards for studies. Blocks were a minimum of two kilometers apart for the influences-on-bird-damage studies (blueberries: mean = 4.6 km apart, SD = 6.5, n = 23 blocks, some sampled in multiple years; grapes: mean = 2.5 km apart, SD = 1.5, n = 19 blocks, some sampled in multiple years). In the case of the tubemen and methyl anthranilate tests, blocks were often closer than two kimometers, because block pairs, or halves of blocks in the case of the methyl anthranilate test, were owned by individual growers.

2.3. Sampling Bird Damage within Blocks

To estimate bird damage within blocks, we divided blocks into four edge strata and one interior stratum, given that the edges of blocks sometimes experience greater damage than interiors [13]. Within a block, edge strata were two rows wide, with the interior stratum comprising all of the other rows. We sampled whole blocks, except in the case of the methyl anthranilate test, where blocks were divided into treatment and control halves. For tests with tubemen, we sampled 2.5-acre areas around the tubemen. These 2.5-acre areas were generally on a corner of the block, so there were only two edge strata and one interior stratum. Since there were only three strata, we increased the number of bushes sampled per stratum to 20 to sample roughly 60 plants, as we did for the influences-on-bird-damage studies.

2.4. Plant Selection

We sampled up to 12 plants per stratum for blueberries, resulting in approximately 60 bushes sampled per block, and up to 20 plants per stratum for grapes, resulting in up to 100 vines sampled per block. We sampled fewer plants in blueberries because of the labor-intensive nature of the sampling (see below). Within each stratum, we randomly selected a starting plant, and then systematically chose the other plants for sampling to provide approximately even coverage of the stratum. For example, if we randomly selected the third plant from the northeast corner as the starting plant for a stratum, and the stratum contained 112 plants in total, we sampled every 10th plant, so that the 11 remaining sample plants were from all of the areas of the stratum.

2.5. Blueberry Cane Selection

For each plant, we randomly selected a number between one and eight; each number represented one of the eight half-winds of the compass rose (NNE, ENE, ESE, SSE, SSW, WSW, WNW, NNW). The half-wind selected represented the side of the plant from which we selected a cane for sampling. We then placed an inconspicuously colored twist tie between 0.5–1 m back from the tip of the cane.

Two to three weeks before the first harvest, we counted all of the green and blue berries from the twist tie toward the tip. Since we needed to return to the same cane to recount the berries, we flagged a plant adjacent to the plant with the selected cane. The twist ties were dark green, and the flagging was placed on the side of the plant opposite the sampled plant; neither of these practices should have influenced berry consumption by birds. During the recount, which happened right before the first harvest, we again counted all of the green and blue berries, and any damaged berries. We subtracted the number of berries counted on the second date from the number of berries counted on the first date, and divided this value by the number of days between the sampling dates to obtain the number of berries per branch lost to birds per day. Our assumption was that the berries that were missing were primarily lost to birds. The growers who we spoke with believed this was a reasonable assumption. If there was a reason to think berries were missing for other reasons, we did not include data from sampled plants. For example, when we returned for a recount, we occasionally found evidence of mammal damage near a sampled plant (e.g., scat on the ground, substantial damage to leaves as well as fruit). Also, for example, damage from hail will leave distinctive marks on the berries, or occasionally it was clear that farm equipment had damaged a plant. If we found evidence of any non-bird damage or loss, these plants were removed from the datasets before analyses.

2.6. Grape Cluster Selection

Bird damage sampling took place right before harvest. We randomly selected one cluster per vine by randomly selecting a number, based on 10-cm intervals, to represent the height of the cluster, and a randomly selected number between one and six to represent the horizontal position of the cluster; three represented the vine stem, zero was the left-hard edge of the vine, six represented the right-hand edge, and the other numbers were equidistant between the edges [13]. Field workers worked in pairs to estimate the percent damage after practice, and used diagrams of clusters with various levels of damage prepared by R.W. Emmett of the Department of Agriculture, Mildura, Victoria, Australia. Damage was estimated to 1% intervals if the damage was less than 10% or more than 90%, and in 5% intervals if it was between 10–90% [13]. If there was evidence of mammal damage near a sampled plant, the plant was removed from the dataset before analyses.

2.7. Tubemen Site Selection in Blueberries

In 2013, we matched four pairs of Bluecrop blocks that had comparable damage levels based on 2012 damage assessments. In 2014, we used blocks of Jersey, and each of our control and treatment block pairs ($n = 5$) were owned by one grower, who stated that the blocks had comparable levels of bird damage. One block per pair in both years was randomly assigned to be a control, and one was a treatment. Treatment blocks had one or two inflatable tubemen "dancing" on the edge and/or interior of the block for two to three weeks before harvest, for approximately 10 h per day. The blocks were 2.6 acres or less with three exceptions: one of 3.1 acres, one of four acres, and one of 5.2 acres. We used tubemen that were approximately 5.5 m tall, with blue, white, and red sections, purchased from LookOurWay®, San Francisco, CA, USA. Tubemen were powered by generators.

2.8. Tubemen Site Selection in Grapes

All of the blocks were Pinot noir. Seven blocks that were assessed for damage levels in 2012 were randomly assigned to tubemen or control treatments in 2014; four of the blocks had tubemen, and three did not. The blocks with tubemen had between two and four tubemen that were moved several times in the weeks before harvest. Sometimes, they were placed on the edge of the blocks, and sometimes, they were placed in the interior. Blocks were 2.2 acres or less in size with two exceptions: one block of 3.6 acres, and one of 4.7 acres.

2.9. Methyl Anthranilate Site Selection in Blueberries

We cooperated with one grower in 2015 who sprayed half of four blocks with Avian Control®
before harvest, according to label directions. The unsprayed halves of the blocks served as controls.
All blocks were Bluecrop. Block sizes were 1.2 acres, 2.6 acres, 3.7 acres, and 20.4 acres.

2.10. Sampling for Fruit-Eating Bird Activity.

For the influences-on-bird-damage studies in blueberries and grapes, we sampled to obtain
estimates of bird abundance using point counts. We performed point counts in blocks in 2012 and
2013, as close in time before harvest as possible, in three regions: Michigan, New York, and the Pacific
Northwest. We only present abundance estimates for Michigan in this manuscript, because of our focus
here on Michigan; estimates for other regions are in Hannay et al. [14]; please see the data analysis
section for more details. In most of the cases, two observers conducted independent point counts
simultaneously, not communicating during or after the count [15]. All of the birds that were visually
detected within a 25-m radius were recorded for 15 min. Each observer's point count was considered
a separate temporal replicate. Point counts were conducted at random positions at both edge and
interior points in blocks where interior points could be located at least 50 m from any edge. Interior
and edge points were considered temporal replicates. Please see Hannay et al. [14] for additional
details about point count sampling.

For the tubemen trials in blueberries, we sampled bird activity with 30-min observation periods,
with the observer positioned 10 m from the southwest corner of the block, unless visibility or other
issues prevented use of this corner, in which case another block corner was used. An observer scanned
a one-acre area of the block with binoculars, and recorded each individual bird that entered the block.
Therefore, these counts represent bird activity rather than numbers of individuals, because the same
individuals could have visited a block more than once within the 30-min period. A few blocks were
sampled more than once. We conducted 13 h of paired observations in blueberries, i.e., with half-hour
observations at the control blocks, and half-hour observations at the paired treatment blocks, usually
within the same day, although some of the paired observations were on consecutive days.

For the methyl anthranilate trial, we conducted bird sampling on one day close to harvest for each
of the pairs of half-blocks. We followed the protocol described above for the tubemen trials, except that
the observer scanned the entire half of the block (not just one acre, as in the tubemen trials) that was
either the control or treatment, while another observer, at the same time, scanned the other half of the
block. Each treatment and control was sampled once, and we had a total of 3.5 h of bird observations
for this trial; one observation had to be shortened to 15 min rather than 30 min for logistical reasons.

We did not conduct bird sampling for the tubemen trial in grapes, but have examined data from
seven hours of bird observations conducted in six vineyards from our other studies to determine which
fruit-eating birds commonly visited vineyards.

A bird species was classified as fruit-eating if it met one of two criteria: (1) it was observed eating
fruit during the observations we conducted for this and previous studies [14], (2) or fruit consumption
was documented in the Birds of North American entry for the species [16].

2.11. Forest Cover Estimations

We classified land cover/use (LCLU) from the NAIP (National Agriculture Imagery Program [17]
one-meter orthoimagery to determine the proportion of forest within roughly 500-m buffers around
each study block. Please see Lindell et al. [4] for details and justification of use of this imagery.
The classification was subsequently verified/corrected based on ground-truth observations by
field workers.

For the influences-on-bird-damage study in blueberries, we created a categorical variable from the
NAIP imagery; low (<50%) or high (50% or greater) forest cover in the 500-m buffer around each block.
We chose the breakpoint as 50%, because the largest gap in the forest cover variable was between

44–54%. For the corresponding study in grapes, we discovered that the surrounding forest cover was less than 50% for all but one block, so instead, we used ground-truth data to create another more local-scale variable related to forest cover, i.e., the number of block edges of deciduous or coniferous forest at least five meters tall. We chose five meters because at that height, woody cover would generally provide resources—such as cover from predators—that could cause fruit-eating birds to move regularly between the woody cover and fruit blocks.

2.12. Data Management and Statistical Analyses

To determine the proportions of bird damage for both blueberry and grape blocks, we calculated weighted estimates by first determining the mean damage per stratum from the 12 (blueberries) or 20 (grapes) plants sampled in each of the strata per block. We then multiplied the mean for each stratum by the proportion of the plants in that stratum, given the number of plants in the whole block, and added the resulting values to arrive at an estimate of the bird damage for the block [4,13].

For the influences-on-bird-damage studies for blueberries and grapes, we used Proc Glimmix [18]. Proc Glimmix provides generalized linear mixed models (models with fixed and random factors), and can be used with data with non-normal distributions [19]. We considered year, grower, and forest variables as fixed factors. Year was coded as a dummy variable. We included site as a random factor, because some sites were sampled in more than one year, and bird damage within a site across years could have been correlated. We used a beta distribution and a logit link. We compared the null models (only the intercept) with all one-variable, two-variable, and three-variable models, and with the one global, four-variable model. We compared the AICc (Akaike information criterion corrected for small sample size) values of the models, and considered all of the models within two AICc units of the model with the lowest AICc value as potential final models. The model selection procedures for blueberries and grapes did not indicate that any interactions would be valuable additions to models. We checked the Pearson chi-square value/df to assess that final models had adequate fit.

For the tubemen test in blueberries, we combined data from 2013 and 2014 for analyses given the similarity of the experimental design, i.e., paired blocks, and used a paired *t*-test to compare bird damage levels between the control and tubemen treatments. For the grape tubemen test, we lost one control block because of unforeseen bird management efforts by the owner, so we conducted an unpaired *t*-test with the remaining treatment and control blocks. For the methyl anthranilate test in blueberries, we used a paired *t*-test to compare treated and control halves of fields. *t*-Tests for the tubemen tests were one-tailed, because we expected that treatment blocks with tubemen would have lower damage. The *t*-test for the methyl anthranilate test was two-tailed to account for the possibility that the chemical application could increase fruit damage. We calculated Hedges' *g* as a measure of effect size for the deterrent tests with tubemen and the methyl anthranilate spray.

We estimated the abundance of fruit-eating birds at point count locations from the point count data using binomial mixture models. These models use site and temporal replicates to estimate abundance [20] and account for factors that may influence the abundance as well as the sampling process. Separate binomial mixture models for fruit-eating bird abundance were constructed for blueberries and grapes. Although we only generated estimates for Michigan, we used data from three regions, Michigan, New York, and the Pacific Northwest, collected in 2012 and 2013, to improve estimates, given the large number of point counts with no bird detections, leading to an overdispersion of the data. Multiple study blocks in each region were used as site replicates. In the cases where two observers conducted independent point counts simultaneously, each point count was used as a separate temporal replicate. When separate point counts were conducted in the edge and interior locations of some blocks, these were also considered temporal replicates. Therefore, there was a possible maximum of four temporal replicates for a study block when both edge and interior points were sampled by two observers. Models were analyzed in a Bayesian framework using the R2jags package [21].

The abundance model is as follows:

$$N_{i,k} \sim \text{Poisson} (\lambda_{i,k}) \tag{1}$$

$$\log(\lambda_{i,k}) = \alpha_k + \beta_1 \times (\text{Region}_i) + e_i, \text{ where } e_i \sim \text{Normal} (0, \sigma^2_\lambda) \tag{2}$$

The observation model is as follows:

$$y_{i,j,k} \mid N_{i,k} \sim \text{Binomial} (N_{i,k}, p_{i,j,k}) \tag{3}$$

$$\text{logit}(p_{i,j,k}) = \beta_k + \delta_{i,j,k}, \text{ where } \delta \sim \text{Normal} (0, \sigma^2_p) \tag{4}$$

Above, k is the number of years in the study (two), j is the number of temporal replicates, and i is the number of study blocks. Estimated site abundance is represented by N. Overdispersion, which is common in count data with many 'zero' counts, was accounted for in the abundance models by including a random variable for block (e_i), and in the observation model through a random variable for each temporal replicate ($\delta_{i,j,k}$). The terms σ^2_λ and σ^2_p represent the standard deviation of λ and p, respectively, and are estimated by the model with uninformative priors. Point count data were collected for two different years, 2012 and 2013, which were modeled with different intercepts (α_1 and α_2). We used uninformative priors for each model. We ran three Markov chains for 350,000 iterations, with the first 50,000 iterations being excluded as 'burn in'. Model convergence was checked by making sure that the Rhat values for all of the models were within 0.1 of 1, which is considered to be an acceptable range for convergence [22], as well as by looking at the mixing of the three Markov chains. We inspected model fit by using the ratio between simulated and actual data; a good 'fit' is around one [20]. Both models had a fit of 1.00 +/− 0.02, and a "Bayesian *p*-value" within 0.02 of 0.5, where values around 0.5 are considered ideal [20]. For additional details about the estimation process, please see Hannay et al. [14].

2.13. Animal Care Statement

These studies were approved by the Michigan State University Institutional Animal Care and Use Committee, approval #04/14-076-00.

3. Results

3.1. Influences on Bird Damage to Blueberries and Grapes

The model selection procedure for the influences-on-bird-damage study for blueberries indicated that the variables yeartwo (2013) and yearthree (2014) were important components of the model with lower damage in 2013 and 2014 compared to 2012 (Table 2 and Figure 1). The model including yeartwo and yearthree had the lowest AICc value, although the model with only yeartwo was within two AICc units, suggesting some support for that model as well [23] (Table 3). Given that the upper and lower confidence limits for the yearthree variable include 0, this variable is likely less important than yeartwo. The final model for grapes included the variable yeartwo (2013) with much lower damage in 2013 compared to 2012 and 2014 (Tables 2 and 4, and Figure 2).

The years were coded as dummy variables; the estimates refer to the years besides the one listed as the explanatory variable. For example, the estimate in the yeartwo row for blueberries is the estimate for yearone (2012) and yearthree (2014) combined, with yeartwo as the reference value (estimate = 0). C.L. refers to confidence limits.

Table 2. Generalized linear model results for influences on bird damage for blueberries and grapes.

Crop	Explanatory Variables in Best Model	Estimate	Standard Error	DF	*t*-Value	*p*-Value	Lower C.L.	Upper C.L.
Blueberries	Yeartwo (2013)	0.856	0.281	10	3.05	0.012	0.173	1.495
	Yearthree (2014)	0.797	0.278	10	2.87	0.017	−0.246	1.787
Grapes	Yeartwo (2013)	1.060	0.438	8	2.42	0.0417	0.051	2.069

C.L.: Confidence limit. DF: Degrees of freedom.

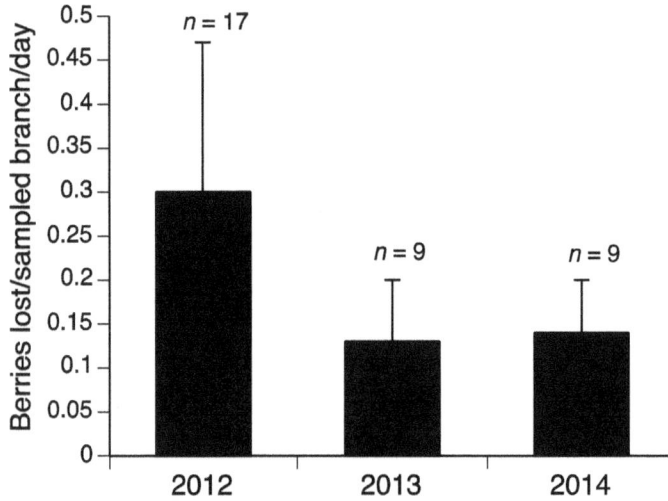

Figure 1. Blueberries lost to birds in Michigan 'Bluecrop' fields. Error bars are standard deviations. *n* = number of blocks sampled.

Table 3. Models compared for influences-on-bird-damage to blueberries.

Model	AICc	Delta AICc	Likelihood	Akaike Weight
Yeartwo Yearthree	−34.27	0	1	0.39
Yeartwo	−33.73	0.54	0.76	0.30
Yeartwo Yearthree Forestpercentcat	−31.5	2.77	0.25	0.10
Yeartwo Forestpercentcat	−31.24	3.03	0.22	0.09
Null	−30.89	3.38	0.18	0.07
Yearthree	−29.3	4.97	0.08	0.03
Forestpercentcat	−28.55	5.72	0.06	0.02
Yearthree Forestpercentcat	−26.59	7.68	0.02	0.01
Yeartwo Yearthree Grower	−18.45	15.82	0.00	0.00
Yeartwo Grower	−16.85	17.42	0.00	0.00
Grower	−16.41	17.86	0.00	0.00
Yearthree Grower	−14.01	20.26	0.00	0.00
Yeartwo Yearthree Grower Forestpercentcat	−13.25	21.02	0.00	0.00
Yeartwo Grower Forestpercentcat	−13.01	21.26	0.00	0.00
Grower Forestpercentcat	−12.6	21.67	0.00	0.00
Yearthree Grower Forestpercentcat	−9.38	24.89	0.00	0.00

AICc: Akaike information criterion corrected for small sample size.

Table 4. Models compared for influences-on-bird-damage to grapes.

Model	AICc	Delta AICc	Likelihood	Akaike Weight
Yeartwo	−85.19	0	1	0.39
Yeartwo Numberforestedges	−82.91	2.28	0.32	0.12
Yeartwo Yearthree	−82.42	2.77	0.25	0.10
Null	−80.95	4.24	0.12	0.05
Numberforestedges	−80.53	4.66	0.10	0.04
Yeartwo Yearthree Numberforestedges	−79.9	5.29	0.07	0.03
Yearthree	−79.01	6.18	0.05	0.02
Yearthree Numberforestedges	−77.97	7.22	0.03	0.01
Yeartwo Grower	−38.98	46.21	0.00	0.00
Grower	−35.05	50.14	0.00	0.00
Yeartwo Grower Numberforestedges	−27.64	57.55	0.00	0.00
Yeartwo Yearthree Grower	−26.59	58.6	0.00	0.00
Yearthree Grower	−25.59	59.6	0.00	0.00
Grower Numberforestedges	−24.71	60.48	0.00	0.00
Yearthree Grower Numberforestedges	−12.83	72.36	0.00	0.00
Yeartwo Yearthree Grower Numberforestedges	−12.04	73.15	0.00	0.00

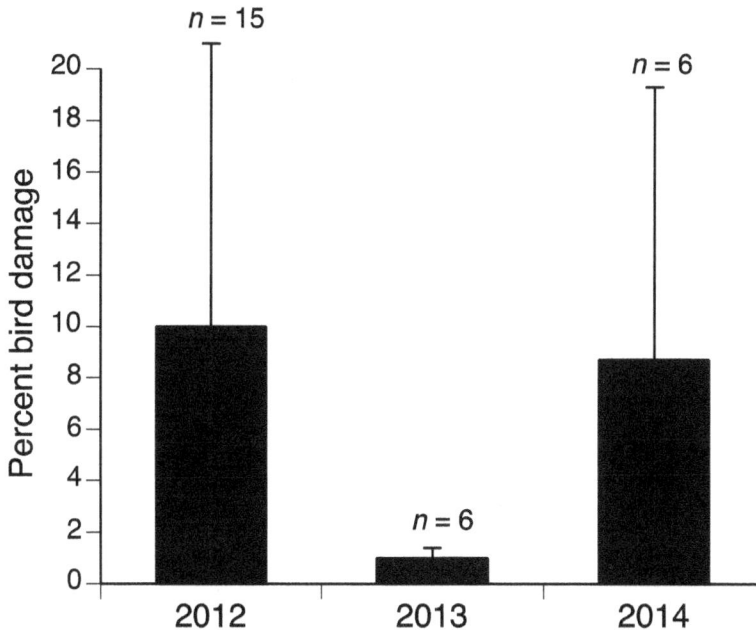

Figure 2. Percent bird damage in Michigan Pinot noir vineyards. Error bars are standard deviations. *n* = number of blocks sampled.

3.2. Bird Deterrent Tests

Bird damage to blueberries was not significantly different in blocks with tubemen compared to paired blocks without tubemen, although there was a non-significant trend toward lower damage in tubemen blocks (paired *t*-test, $t = 1.64$, df = 8, $p = 0.07$, Hedges' *g* = 0.65; CI = −0.23–1.65, Figure 3). Three pairs of blocks showed at least 15% lower damage in treatment blocks; however, there was enough variation in outcomes in the other blocks to render the overall comparison between treatments and controls not significantly different. Similarly, grape blocks with tubemen did not have bird damage levels that were significantly different from the blocks that lacked tubemen (*t*-test, $t = 0.25$, df = 5,

$p = 0.41$, Hedges' $g = 0.16$; CI = -1.32–1.68; mean percent bird damage in control blocks was 4.7 ± 2.1 SD and in tubemen blocks was 4.3 ± 1.5 SD).

Bird damage was not significantly different in the halves of blocks treated with the methyl anthranilate product, Avian Control®, compared to halves that were not treated (paired *t*-test, $t = 0.83$, df = 3, $p = 0.47$, Hedges' $g = 0.49$; CI = -1.04–2.29, Figure 4).

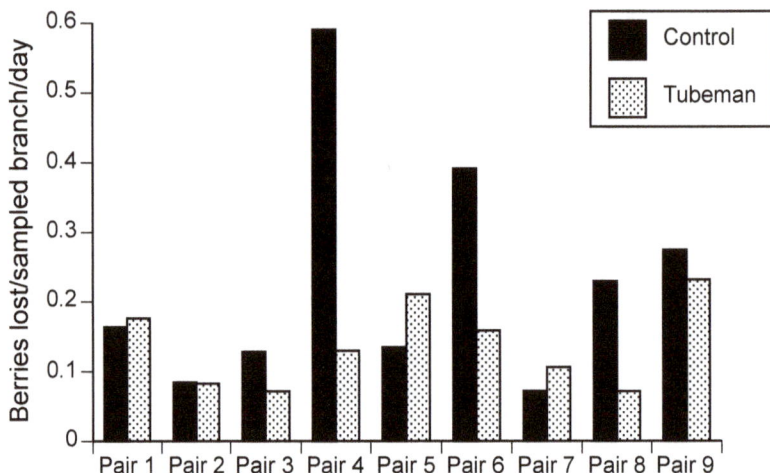

Figure 3. Blueberries per sampled branch per day lost to birds in control and tubemen blocks, 2013 and 2014.

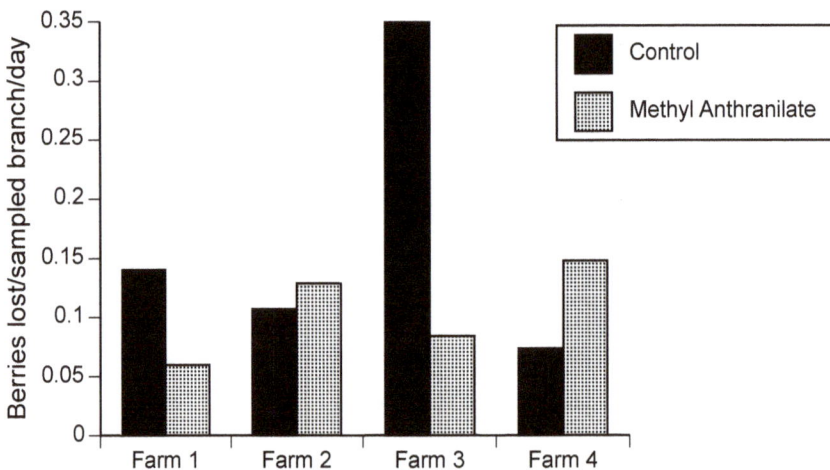

Figure 4. Blueberries per sampled branch per day lost to birds in control halves of fields compared to halves of fields treated with Avian Control® in 2015.

3.3. Fruit-Eating Bird Abundance and Activity.

The large degree of overlap of the credible intervals for 2012 and 2013 in both blueberries and grapes indicates that fruit-eating bird abundance per count area did not differ between the years (Table 5). The standard deviations were large, given the great variability in bird detections for the point counts, with many zeros. Thus, the estimates are imprecise with wide credible intervals.

Table 5. Estimates of fruit-eating bird abundance in Michigan blueberries and grapes. Means are per point count area.

Crop	Year	Mean	Standard Deviation	Credible Intervals 2.5%	97.5%
Blueberries	2012	22.12	14.61	10.24	54.95
	2013	9.75	27.10	1.35	57.83
Grapes	2012	1.88	4.46	0.13	11.01
	2013	4.30	9.22	0.27	27.34

Fruit-eating bird activity was not significantly different in paired blueberry blocks with and without tubemen (paired t-test, $t = 0.54$, df = 12, $p = 0.60$). The mean bird visits to control blocks was 30.8 (SD = 55.4), while the mean bird visits to tubemen blocks was 20.3 (SD = 42.7). We could identify the bird to species for 606 of 663 visits (91.4%). Of the visits by identified species, the four most common fruit-eating species in the control blocks were, from most common, European starlings (*Sturnus vulgaris*, 40.7% of identified visits), American robins (*Turdus migratorius*, 22.4% of identified visits), common grackles (*Quiscula quiscula*, 18.1% of identified visits), and song sparrows (*Melospiza melodia*, 7.1% of identified visits). For the tubemen blocks, the four most common fruit-eating species were European starlings (70.9% of identified visits), American robins (10.8% of identified visits), song sparrows (5.6% of identified visits), and American goldfinches (*Spinus tristis*, 5.2% of identified visits).

Fruit-eating bird activity was not significantly different in halves of blueberry blocks with control halves and halves treated with a methyl anthranilate spray (paired t-test, $t = 0.70$, df = 3, $p = 0.53$). Of 51 visits by birds, 42 (82.4%) were identified to species. The three most commonly identified species in the control halves of blocks were northern flickers (*Colaptes auratus*, 25.0% of identified visits), tufted titmice (*Baeolophus bicolor*, 25% of identified visits), and American robins (18.8% of identified visits). The only two identified species visiting the treatment halves were American robins and a brown thrasher (*Toxostoma rufum*).

Sixty-four of 73 visits of fruit-eating birds to Pinot noir blocks over seven hours of observation were by American robins, with a few other species comprising the remainder of the visits.

4. Discussion

This work corresponds with past work [4] demonstrating that the percent bird damage can vary greatly from year to year in fruit crops, which is likely because bird consumption remains relatively consistent over time, while the availability of fruit varies. In our previous work with tree fruits, we found remarkably consistent numbers of sweet cherries either lost to or damaged by birds, despite large differences in sweet cherry abundance from year to year (Table 6 in Lindell et al. [4]). Thus, in high-yield years, bird damage is diluted over a greater abundance of fruit, resulting in lower proportions of damage. In 2012, Michigan blueberry fields yielded 39 million kg; in 2013, this figure was 53 million kg, and in 2014, it was 45 million kg [24] (p. 35). The bird damage patterns reflect this varying abundance of fruit over the years with a lower percent damage in 2013 and 2014 and the highest in 2012 (Figure 1). The pattern was similar in grapes; Michigan grape production was nearly 35 million kg in 2012, compared to 85 million kg in 2013 and 57 million kg in 2014 [24] (p. 37). Bird damage results from these years show lower percent damage in 2013 compared to the other lower-yield years (Figure 2). Work in other crops reinforces the general principle that greater crop abundance results in lower percent damage by vertebrate pests; in a year with high food abundance, including crops and natural forage, percent damage to sunflowers by deer and elk was low; the following year, with reduced food abundance, percent crop damage increased [25]. Generally, absolute pest damage in a region should be consistent from year to year, while local damage depends on the amount of crop available to pests [26]. For example, where corn and sunflower covered the largest acreage in North Dakota, bird damage was lowest [27].

Our fruit-eating bird abundance estimates provided some evidence for relatively consistent bird numbers from year to year; the credible intervals for estimates for 2012 and 2013 broadly overlapped, indicating no differences in bird numbers between years. However, the credible intervals were large, reflecting the overdispersed data, i.e., a large number of point counts with no bird detections. We return to this point at the end of the discussion. We also note that fruit-eating bird abundance may not be as consistent from year to year in other regions where irruptive species such as the Bohemian waxwing, *Bombycilla garrulus*, are important fruit consumers.

The lack of effect of grower on percent bird damage is somewhat surprising in that growers manage their fields differently, based on experience and resources. We also anticipated that forest cover near blocks would influence percent damage, as we found in sweet and tart cherries [4]. However, the year effect is apparently large enough to outweigh any grower and forest land-cover effects.

Based on these results and previous work, growers should be able to predict when and where fruit is most likely to be at risk from bird damage. Times and areas where fruit abundance is low are likely to be high-risk situations for fruit. For example, orchards with varieties of fruit that ripen early in the season will provide some of the few food sources available for fruit-eating birds, and thus be particularly susceptible to bird damage (e.g., Eaton et al. [28]). Similarly, edges of fruit blocks adjacent to non-fruit landcovers will be at higher risk than edges adjacent to other fruit blocks [4]. If growers know it is likely to be a low-yield year, a higher proportion of their crop is likely to be taken by birds, and so they should prepare to invest more in bird management.

Fruit growers believe that many bird deterrent techniques are only slightly or moderately effective in deterring birds. The exception was netting, which was viewed by over 50% of growers surveyed as very effective [5]. The present work indicates that inflatable tubemen do not consistently reduce fruit damage. The comparison of bird damage per 2.5-acre sampling area in blueberries with and without tubemen nearly reached the level of statistical significance ($p < 0.07$), indicating that tubemen may reduce damage in some contexts (see also Steensma et al. [29]). Although the tubemen move somewhat randomly, it is likely that birds habituate over time to their presence, as is the case with scare tactics generally (e.g., Cook et al. [30] and Summers [31]). If growers use this technique, we recommend that they use several tubemen for each five acres, place them on the edges of blocks, begin running them before the fruit ripen, and change their positions every few days.

Similarly, the test with a methyl anthranilate spray did not result in detectable differences in bird damage between the control and treatment halves of fields. Systematic tests of methyl anthranilate sprays in fruit have generally not shown strong evidence of efficacy against birds (e.g., Avery et al. [11]), which is potentially because any irritation caused by methyl anthranilate sprays is short-term and birds can move away from treated fruit; thus, they do not learn to avoid it, because they are not exposed to large dosages [32].

Several other bird deterrent techniques that have not been tested in this study show promise; some only show promise in particular contexts. Natural predators can be attracted to fruit-production regions through landscape enhancements, such as for example nest boxes [33]. We demonstrated that box-nesting America kestrels, *Falco sparverius*, can be attracted to sweet cherry orchards and reduce the abundance of fruit-eating birds [34]. Some varieties of blueberries will ripen during the kestrel incubation and nesting periods, which is when the kestrels are most likely to provide protection from pest birds; as a result, this technique may be useful in some blueberry production regions. Grapes ripen after the kestrel nesting period, so this strategy is not well-suited for grapes. Deterrence by natural predators has the additional advantage that it is preferred by consumers compared to other deterrent techniques that are deemed less "natural", and so growers may be able to garner better prices if they advertise their use of natural predators [35,36].

Unmanned aerial systems (UAS) may also be valuable bird deterrent systems, although work to date is very limited [37]. Since UAS are not simply passive scare devices, but could actually chase fruit-eating birds (while respecting wildlife regulations), they may be more effective as deterrents.

Other recent developments in bird management focus on disrupting input to birds' sensory systems, which is likely to be a fruitful direction for future research. "Sonic nets", for example, broadcast noise that interferes with birds' communication channels, and have shown effectiveness in deterring birds from airfields [38]. This technique may be useful in fruit crops, although no systematic studies have yet been performed. Preliminary studies of laser scarecrows, where a laser beam sweeps over a field, showed apparent reduced bird damage in sweet corn fields with the devices compared to those without, although no statistical evidence was presented [39]. By imposing an actual cost on birds through reducing their ability to detect predators and/or communicate with conspecifics, these deterrent techniques may be less susceptible to habituation than deterrent techniques with no cost. A recent review discusses conservation and management strategies for seabirds focused on bird sensory systems that suggests some modalities (for example, olfaction) and life history traits (for example, coloniality) that could be fruitful targets for research into bird deterrence [40]. In all cases where birds could potentially cause significant crop damage, we recommend that growers consider an integrated pest management approach, using several deterrent techniques. We also suggest growers prepare and employ a management plan that is specific to their fields and considers risk factors for bird damage and potential mitigation strategies [3] (pp. 211–218).

We end with two methodological recommendations. Bird activity is often quite variable from place to place and time to time. A large number of our point counts resulted in no bird detections. This overdispersion of data can result in large credible intervals for bird abundance estimates. Thus, it is challenging to capture bird abundance, activity, and/or species composition in "snapshot"-type sampling, such as 10 or 15-min point counts. For future work, we suggest longer observation periods, such as the 30-min periods we used in concert with deterrent tests, over standardized areas. This type of sampling is likely to result in more accurate measures of bird abundance, activity, and species composition. Second, the effect sizes for our bird deterrent tests tended to have large confidence intervals, despite our having a greater degree of site replication than many previous studies. We suggest that before-and-after sampling (sampling birds or damage before and after deployment of a deterrence strategy) would be a helpful addition to pairs of treatment and control sites for tests of bird deterrence. Since birds are mobile and cover large areas, treatment and control sites should be relatively large, limiting the number of sites that can be sampled. Before-and-after sampling would have to be conducted over shorter time scales, for example over two-day periods very close to harvest, to ensure some confidence that conditions, including the ripeness of the fruit, are similar before and after the deployment of the strategy. Before-and-after sampling may allow sampling of a larger number of blocks, with environmental conditions controlled, thus increasing sample sizes and increasing the level of confidence in the results.

5. Conclusions

Reducing bird damage to fruit remains a challenging issue. We found that the year was the most important variable in explaining bird damage to blueberries and grapes, with high-yield years having lower proportions of bird damage. Thus, in high-yield years, growers may not view bird management as a priority. Two deterrent strategies, inflatable tubemen and a spray containing methyl anthranilate, did not consistently reduce bird activity or damage. Attracting natural predators with nest boxes and disrupting birds' sensory systems are two promising bird deterrent strategies that should continue to be investigated. Given birds' mobility and variability in activity over time and space, before-and-after sampling, along with pairs of treatment and control blocks, should be considered to increase sample sizes in deterrent tests.

Author Contributions: C.A.L. conceived and designed the experiments; B.C.H. and C.A.L. performed the field work; C.A.L. and M.B.H. analyzed the data; C.A.L. wrote the paper.

Funding: This work was supported by the Specialty Crop Research Initiative of USDA/NIFA (Grant number: 2011-51181-30860).

Acknowledgments: We thank the many fruit growers who provided access to their fields, assisted with tests, and pulled us out of the mud when necessary. We are grateful for the many and varied contributions of Shayna Wieferich, Della Fetzer, and Emily Oja. We thank the Stone Soap Company for providing Avian Control®.

Conflicts of Interest: The authors declare no conflict of interest. The founding sponsors had no role in the design of the study; in the collection, analyses, or interpretation of data; in the writing of the manuscript, and in the decision to publish the results.

References

1. National Agricultural Statistics Service (NASS). *Noncitrus Fruits and Nuts 2005 Summary*; Fr Nt 1–3(06); U.S. Department of Agriculture: Washington, DC, USA, 2006.
2. National Agricultural Statistics Service (NASS). *Noncitrus Fruits and Nuts 2016 Summary*; U.S. Department of Agriculture: Washington, DC, USA, 2017; ISSN 1948-2698.
3. Tracey, J.; Bomford, M.; Hart, Q.; Saunders, G.; Sinclair, R. *Managing Bird Damage to Fruit and Other Horticultural Crops*; Bureau of Rural Sciences: Canberra, Australia, 2007. Available online: https://www.dpi.nsw.gov.au/content/_data/assets/pdf_file/0005/193739/managing_bird_damage-full-version.pdf (accessed on 25 September 2018).
4. Lindell, C.A.; Steensma, K.M.M.; Curtis, P.D.; Boulanger, J.R.; Carroll, J.E.; Burrows, C.; Lusch, D.P.; Rothwell, N.L.; Wieferich, S.L.; Henrichs, H.M.; et al. Proportions of bird damage in tree fruits are higher in low-fruit-abundance contexts. *Crop Prot.* **2016**, *90*, 40–48. [CrossRef]
5. Anderson, A.; Lindell, C.A.; Moxcey, K.M.; Siemer, W.F.; Linz, G.M.; Curtis, P.D.; Carroll, J.E.; Burrows, C.L.; Boulanger, J.R.; Steensma, K.M.M.; et al. Bird Damage to Select Fruit Crops: The cost of damage and the benefits of control in five states. *Crop Prot.* **2013**, *52*, 103–109. [CrossRef]
6. Lock, K.; Pomerleau, J.; Causer, L.; Altmann, D.R.; McKee, M. The global burden of disease attributable to low consumption of fruit and vegetables: Implications for the global strategy on diet. *Bull. World Health Organ.* **2005**, *83*, 100–108. [PubMed]
7. World Health Organization. Global Strategy on Diet, Physical Activity and Health. Available online: http://www.who.int/nmh/wha/59/dpas/en/ (accessed on 25 September 2018).
8. Johnson, D.B.; Guthery, F.S.; Koerth, N.E. Grackle damage to grapefruit in the lower Rio Grande Valley. *Wildl. Soc. Bull.* **1989**, *17*, 46–50.
9. Otis, D.L.; Kilburn, C.M. Influence of environmental factors on blackbird damage to sunflower. In *Fish Wildl.*; Tech. Rep. 16; U.S. Fish and Wildlife Service: Washington, DC, USA, 1988.
10. Mason, J.R.; Adams, M.A.; Clark, L. Anthranilate repellency to starlings: Chemical correlates and sensory perception. *J. Wild. Manag.* **1989**, *53*, 55–64. [CrossRef]
11. Avery, M.L.; Primus, T.M.; DeFrancesco, J.; Cummings, J.L.; Decker, D.G.; Humphrey, J.S.; Davis, J.E.; Deacon, R. Field evaluation of methyl anthranilate for deterring birds eating blueberries. *J. Wild. Manag.* **1996**, *60*, 929–934. [CrossRef]
12. Dieter, C.D.; Warner, C.S.; Curiong, R. Evaluation of foliar sprays to reduce crop damage by Canada geese. *Hum. Wildl. Interact.* **2014**, *8*, 139–149. Available online: https://digitalcommons.usu.edu/hwi/vol8/iss1/15 (accessed on 25 September 2018).
13. Tracey, J.; Saunders, G.R. A technique to estimate bird damage in wine grapes. *Crop Prot.* **2010**, *29*, 435–439. [CrossRef]
14. Hannay, M.B.; Boulanger, J.R.; Curtis, P.D.; Eaton, R.A.; Hawes, B.C.; Leigh, D.K.; Rossetti, C.A.; Steensma, K.M.M.; Lindell, C.A. Bird species and abundances in fruit crops and implications for bird management. *Crop Prot.*, in review.
15. Nichols, J.D.; Hines, J.E.; Sauer, J.R.; Fallon, F.W.; Fallow, J.E.; Heglund, P.J. A double-observer approach for estimating detection probability and abundance from point counts. *Auk* **2000**, *117*, 393–408. [CrossRef]
16. Rodewald, P. *The Birds of North America*; Cornell Laboratory of Ornithology: Ithaca, NY, USA, 2015; Available online: http://bna.birds.cornell.edu/BNA/ (accessed on 1 January 2018).

17. National Agricultural Imagery Program (NAIP). Information Sheet. United States Department of Agriculture Farm Service Agency. Available online: http://fsa.usda.gov/Internet/FSA_File/naip_info_sheet_2013.pdf (accessed on 13 August 2014).
18. SAS Institute Inc. SAS 9.3. for Windows. Cary, NC, USA, 2002–2010. Available online: http://support.sas.com/software/93/ (accessed on 24 May 2018).
19. SAS Institute Inc. SAS/STAT® 9.3 User's Guide. Available online: https://support.sas.com/documentation/cdl/en/statug/63962/HTML/default/viewer.htm#titlepage.htm (accessed on 24 May 2018).
20. Kéry, M.; Schaub, M. *Bayesian Population Analysis Using WinBUGS, a Hierarchical Perspective*; Academic Press: Waltham, MA, USA, 2012; ISBN 10:0123870208.
21. Su, Y.S.; Yajima, M. R2jags: A Package for Running Jags from R. Available online: https://CRAN.R-project.org/package=R2jags (accessed on 1 January 2014).
22. Kéry, M. *Introduction to WinBUGS for Ecologists: Bayesian Approach to Regression, ANOVA, Mixed Models and Related Analyses*; Academic Press: Orlando, FL, USA, 2010; ISBN 10:0123786053.
23. Burnham, K.P.; Anderson, D.R. *Model Selection and Multimodel Inference: A Practical Information-Theoretic Approach*, 2nd ed.; Springer-Verlag: New York, NY, USA, 2002; ISBN 0-387-95364-7.
24. National Agricultural Statistics Service (NASS). Annual Statistical Bulletin. Michigan Agricultural Statistics 2016–2017. Available online: https://www.nass.usda.gov/Statistics_by_State/Michigan/Publications/Annual_Statistical_Bulletin/stats17/agstat17.pdf. (accessed on 25 September 2018).
25. Johnson, H.E.; Fischer, J.W.; Hammond, M.; Dorsey, P.D.; Walter, W.D.; Anderson, C.; VerCauteren, K.C. Evaluation of techniques to reduce deer and elk damage to agricultural crops. *Wildl. Soc. Bull.* **2014**, *38*, 358–365. [CrossRef]
26. Leitch, J.A.; Linz, G.M.; Baltezore, J.F. Economics of cattail (*Typha* spp.) control to reduce blackbird damage to sunflower. *Agr. Ecosyst. Environ.* **1997**, *65*, 141–149. [CrossRef]
27. Klosterman, M.E.; Linz, G.M.; Slowik, A.A.; Homan, H.J. Comparisons between blackbird damage to corn and sunflower in North Dakota. *Crop Prot.* **2013**, *53*, 1–5. [CrossRef]
28. Eaton, R.A.; Lindell, C.A.; Homan, H.J.; Linz, G.M.; Maurer, B.A. American Robins (*Turdus migratorius*) and Cedar Waxwings (*Bombycilla cedrorum*) vary in use of cultivated cherry orchards. *Wilson J. Ornithol.* **2016**, *128*, 97–107. [CrossRef]
29. Steensma, K.; Lindell, C.; Leigh, D.; Burrows, C.; Wieferich, S.; Zwamborn, E. Bird damage to fruit crops: A comparison of several visual deterrent techniques. In Proceedings of the 27th Vertebrate Pest Conference, Newport Beach, CA, USA, 7–10 March 2016; Timm, R.M., Baldwin, R.A., Eds.; University of California: Davis, CA, USA; pp. 196–203.
30. Cook, A.; Rushton, S.; Allan, J.; Baxter, A. An evaluation of techniques to control problem bird species on landfill sites. *Environ. Manag.* **2008**, *41*, 834–843. [CrossRef] [PubMed]
31. Summers, R.W. The effect of scarers on the presence of starlings (*Sturnus vulgaris*) in cherry orchards. *Crop Prot.* **1985**, *4*, 520–528. [CrossRef]
32. Sayre, R.W.; Clark, L. Effect of primary and secondary repellents on European starlings: An initial assessment. *J. Wildl. Manag.* **2001**, *65*, 461–469. [CrossRef]
33. Lindell, C.A.; Eaton, R.A.; Howard, P.H.; Roels, S.M.; Shave, M.E. Enhancing agricultural landscapes to increase crop pest reduction by vertebrates. *Agric. Ecosyst. Environ.* **2018**, *257*, 1–11. [CrossRef]
34. Shave, M.E.; Shwiff, S.A.; Elser, J.L.; Lindell, C.A. Falcons using orchard nest boxes reduce fruit-eating bird abundances and provide economic benefits for a fruit-growing region. *J. Appl. Ecol.* **2018**, *55*, 2451–2460. [CrossRef]
35. Herrnstadt, Z.; Howard, P.H.; Oh, C.-O.; Lindell, C.A. Consumer preferences for 'natural' agricultural practices: Assessing methods to manage bird pests. *Renew. Agric. Food Syst.* **2016**, *6*, 516–523. [CrossRef]
36. Oh, C.-O.; Herrnstadt, Z.; Howard, P. Consumer willingness to pay for bird management practices in fruit crops. *Agroecol. Sust. Food* **2015**, *39*, 782–797. [CrossRef]
37. Mulero-Pazmany, M.; Jenni-Eiermann, S.; Strebel, N.; Sattler, T.; Negro, J.J.; Tablado, Z. Unmanned aircraft systems as a new source of disturbance for wildlife: A systematic review. *PLoS ONE* **2017**, *12*, e0178448. [CrossRef] [PubMed]

38. Swaddle, J.P.; Moseley, D.L.; Hinder, M.K.; Smith, E.P. A sonic net excludes birds from an airfield: Implications for reducing bird strike and crop losses. *Ecol. Appl.* **2016**, *26*, 339–345. [CrossRef] [PubMed]

39. Brown, R. Laser Scarecrows: Gimmick or Solution? University of Rhode Island Vegetable Production Research Reports. Available online: http://digitalcommons.uri.edu/riaes_bulletin/25 (accessed on 30 October 2018).

40. Friesen, M.; Beggs, J.R.; Gaskett, A.C. Sensory-based conservation of seabirds: A review of management strategies and animal behaviours that facilitate success. *Biol. Rev.* **2017**, *92*, 1769–1784. [CrossRef] [PubMed]

agronomy

MDPI

Article

Ergonomic Evaluation of Current Advancements in Blueberry Harvesting

Eunsik Kim [1],*, Andris Freivalds [2], Fumiomi Takeda [3] and Changying Li [4]

[1] Department of Industrial & Manufacturing Systems Engineering, University of Windsor, Windsor, ON N9B 3P4, Canada

[2] Harold and Inge Marcus Department of Industrial and Manufacturing Engineering, The Pennsylvania State University, State College, PA 16802, USA; axf@engr.psu.edu

[3] United States Department of Agriculture—Agricultural Research Service, Appalachian Fruit Research Station, Kearneysville, WV 25430, USA; fumi.takeda@ars.usda.gov

[4] College of Engineering, University of Georgia, Athens, GA 30602, USA; cyli@uga.edu

* Correspondence: eskim@uwindsor.ca; Tel.: +1-519-253-3000 (ext. 5409)

Received: 23 October 2018; Accepted: 14 November 2018; Published: 17 November 2018

Abstract: Work-related musculoskeletal disorders (MSDs) accounted for 32% of days-away-from-work cases in private industry in 2016. Several factors have been associated with MSDs, such as repetitive motion, excessive force, awkward and/or sustained postures, and prolonged sitting and standing, all of which are required in farm workers' labor. While numerous epidemiological studies on the prevention of MSDs in agriculture have been conducted, an ergonomics evaluation of blueberry harvesting has not yet been systematically performed. The purpose of this study was to investigate the risk factors of MSDs for several types of blueberry harvesting (hand harvesting, semi-mechanical harvesting with hand-held shakers, and over-the-row machines) in terms of workers' postural loads and self-reported discomfort using ergonomics intervention techniques. Five field studies in the western region of the United States between 2017 and 2018 were conducted using the Borg CR10 scale, electromyography (EMG), Rapid Upper Limb Assessment (RULA), the Cumulative Trauma Disorders (CTD) index, and the NIOSH (National Institute for Occupational Safety and Health) lifting equation. In evaluating the workloads of picking and moving blueberries by hand, semi-mechanical harvesting with hand-held shakers, and completely mechanized harvesting, only EMG and the NIOSH lifting equation were used, as labor for this system is limited to loading empty lugs and unloading full lugs. Based on the results, we conclude that working on the fully mechanized harvester would be the best approach to minimizing worker loading and fatigue. This is because the total component ratio of postures in hand harvesting with a RULA score equal to or greater than 5 was 69%, indicating that more than half of the postures were high risk for shoulder pain. For the semi-mechanical harvesting, the biggest problem with the shakers is the vibration, which can cause fatigue and various risks to workers, especially in the upper limbs. However, it would be challenging for small- and medium-sized blueberry farms to purchase automated harvesters due to their high cost. Thus, collaborative efforts among health and safety professionals, engineers, social scientists, and ergonomists are needed to provide effective ergonomic interventions.

Keywords: blueberry harvesting; work-related musculoskeletal disorders; ergonomics intervention

---

1. Introduction

Work-related musculoskeletal disorders (MSDs) accounted for 32% (285,950 cases) of days-away-from-work cases in private industry in 2016 and occurred at a rate of 29.4 cases per 10,000 full-time equivalent workers [1]. The term musculoskeletal disorder refers to injuries and disorders of the locomotor apparatus, i.e., muscles, tendons, bones, cartilage, blood vessels, ligaments, and nerves. Work-related musculoskeletal disorders (MSDs) include all musculoskeletal disorders that are induced or aggravated by work and the circumstances of its performance. These painful and often disabling injuries generally develop gradually over weeks, months, and years. According to the National Institute for Occupational Safety and Health (NIOSH), several epidemiological studies have demonstrated evidence of a causal relationship between physical exertion at work and work-related musculoskeletal disorders (MSDs) [2].

MSDs have been associated with repetitive motion, excessive force, awkward and/or sustained postures, and prolonged sitting and standing, all of which are required in farm workers' labor. Furthermore, workers are often paid on a piece-rate system, providing an incentive to work at high speed and skip recommended breaks. This results in labor-intensive practices and high rates of musculoskeletal disorders among farmers and farm workers [3–7].

While numerous epidemiological studies on the prevention of MSDs in Agriculture have been conducted [4,6,8–10], an ergonomics evaluation of manual blueberry harvesting has not yet been systematically performed. The highbush blueberry industry in the United States has experienced rapid growth in the past three decades, producing 231 megatons of blueberries per year, accounting for about 57% of the annual production of highbush blueberries worldwide [11]. Furthermore, as consumers become increasingly aware of healthful eating and of blueberries' convenience, flavor, and ease of consumption in various snacks, salads, and baking, North America, including the U.S., Canada, and Mexico, will continue to have the most developed fresh blueberry market in the world, as shown in Figure 1.

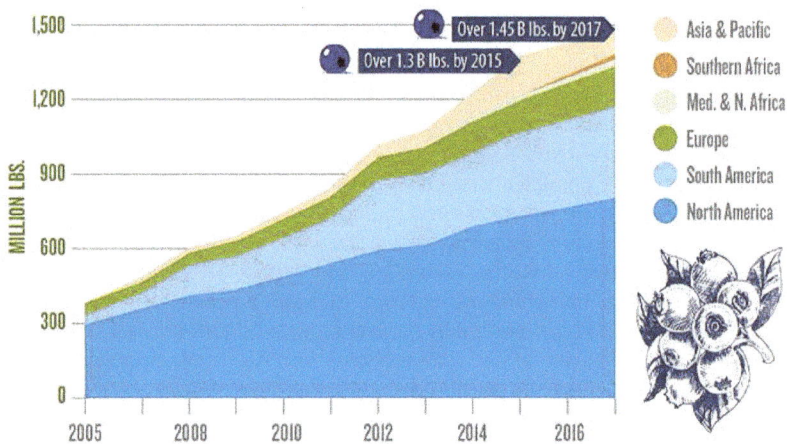

Figure 1. World highbush blueberry production: Growth predictions (source: Cort Brazelton—North American Blueberry Council 2014. Reproduced with permission from North American Blueberry Council).

With such rapid growth in the production and consumption of blueberries in the last 10 years, there has been an increased need for laborers to maintain the bushes and harvest the fruit. These workers are exposed to a high musculoskeletal workload caused by weight, work posture, and repetitive motion. Although over-the-row (OTR) mechanical harvesters are available on the market, much of the blueberry harvesting is still done by hand, since the cost of OTR machines is too high for small- and medium-sized blueberry farms, and since blueberries harvested with OTR machines are likely to be bruised or damaged, making them un acceptable for extended cold storage and the long transport to distant consumers [12,13].

Previous studies on the harvesting of blueberries have focused on harvest efficiency and the quality and quantity of blueberries rather than the labor conditions of the farm worker. In order to improve harvest efficiency, several researchers have sought to develop and evaluate new methods for harvesting blueberries by comparing hand harvesting and semi-machine harvesting with hand-held shakers and OTR machines [11,14,15]. An ergonomic analysis of new machines for mechanical harvesting with a lower cost option (e.g., harvesting blueberries from the ground with hand-held pneumatic shakers and catching fruit using a portable catching frame) was recently described by Takeda et al. [11]. The purpose of this study was to investigate the risk factors of MSDs for each type of blueberry harvesting method (hand-picking, semi-mechanical harvesting, and handling of trays on fully mechanized harvesters) in terms of workers' postural loads and self-reported discomfort using ergonomics intervention techniques.

2. Materials and Methods

We conducted five field studies in three western states (California, Oregon, and Washington) of the United States between 2017 and 2018. Since some companies only permitted data collection via video recording and direct observation, we did not have access to such detailed demographic information as age, education level, smoking status, work experience, and anthropometry data for all participating workers. For hand harvesting, five measurements were used to evaluate workload. First, muscle activity was measured using electromyography (EMG) (Thought Technology Ltd, Montreal West, QC, Canada). The electrode sensors were attached to participants' arms, shoulders, and lower back (Figure 2). Second, Rapid Upper Limb Assessment (RULA) was conducted to estimate the postural risks of work-related upper limb injury. Third, subjective perceived exertion was measured with a Borg CR10 scale to evaluate the workload of harvesting blueberries on the neck, shoulder, arm, hand, and low back. Fourth, the NIOSH lifting equation was used to calculate the risk factor of lifting. Finally, the Cumulative Trauma Disorders (CTD) index was used to compute the potential risk for cumulative trauma disorders caused by repetitive movement. For semi-machine harvesting with hand-held shakers, four measurements were used to evaluate workloads, including EMG, RULA, CTD index, and the NIOSH lifting equation. Lastly, for the completely mechanized harvesting system, only EMG and the NIOSH lifting equation were used, as labor for this system is limited to loading empty lugs and unloading full lugs.

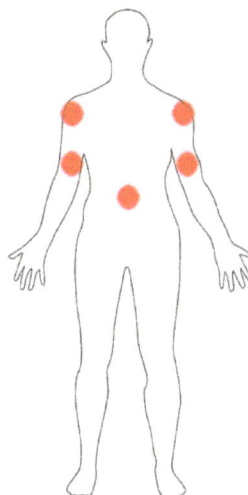

Figure 2. Muscle activity measurement map.

2.1. Blueberry Harveating Method

Highbush blueberries destined for the fresh market have been harvested by hand to maximize quality and postharvest shelf-life. In contrast, nearly all blueberries destined for the process and frozen market is harvested by over-the-row (OTR) machines such as the Oxbo 7440 berry harvester (Oxbo International, Lynden, WA, USA). However, many growers have reported difficulty obtaining sufficient labor for hand harvest operations and the cost of labor is steadily increasing. As a result, some blueberries sold on the fresh market are now harvested by OTR machines, in which fruit is detached with fully mechanized shakers and the workers on these machines handle flats, fill them with berries rolling off the conveyor belt, and then stack filled flats onto a pallet. The third option for harvesting blueberries is to use hand-held, pneumatic, and electric shakers and collect detached berries using a portable catch frame [11] or mechanized fruit conveyance system.

2.1.1. Hand Harvesting

Blueberries do not ripen simultaneously in the cluster. The workers harvesting blueberries by hand wear a shoulder or waist belt harness on which a pail or bucket can be hooked. When one or more berries in a cluster become ripe, the workers clasp the cluster with one or both hands with the palm of the hand underneath the cluster. Ripe berries are teased off the cluster by rolling the thumb or the index finger over them individually until they detach from the pedicel. When several berries are collected in the palms of the hands, the fruit is gently placed in the container. A worker normally picks about 25–30 kg per hour, and expert pickers pick up to 50 kg per hour and work about 6–7 h per day. A flat or pail of blueberries weighs about 7 to 10 kg, depending on the size of box. Thus, each worker picks between 20 and 50 pails per day. When the containers are filled with blueberries, the workers lift, lower, or carry them to the weighing station located at the edge of the field. Depending on the length of the row, workers walk out of the field carrying three or four containers, each weighing between 7 and 10 kg, for as much as 200 m (see Figure 3). Once the fruit is weighed and transferred to flats, workers carry the empty picking containers back through the row to harvest more blueberries. Examples of carrying postures are shown in Figure 3. The total number of carrying trips can be between 10 and 20 times per day for each worker.

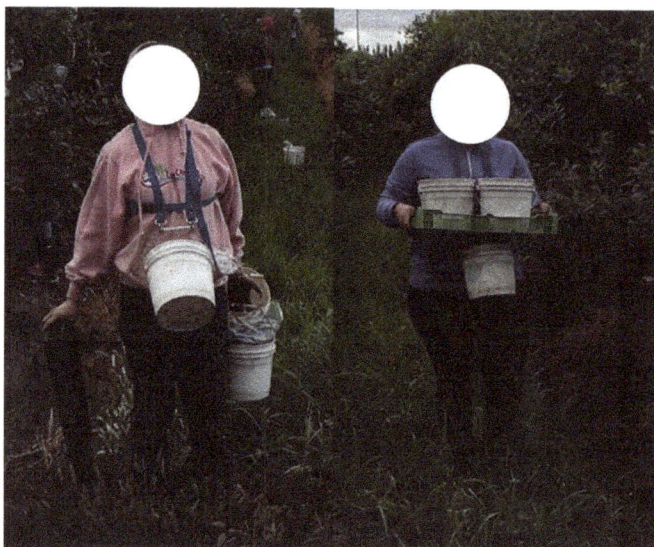

Figure 3. Examples of carrying posture.

2.1.2. Semi-Mechanical Harvesting with Hand-Held Shakers

For the semi-mechanical harvesting method, a long-handled, pneumatically-operated olive harvester (Campagnola, Bologna, Italy) was used to harvest blueberries from a platform on a modified over-the-row harvester (Model 7240, Oxbo International, Lynden, WA, USA). The olive harvester (10) consisted of a 1.2-m-long aluminum tube with a grip in the mid-section and the trigger handle grip at the end. At the other end of the tube, an attachment consisting of an air motor/piston housing to which two reciprocally moving heads with four 20-cm-long plastic tines were mounted. The tips of tines had a lateral displacement of about 5 cm. The speed at which the air motor rotated and actuated the piston's in-and-out motion, which caused the two heads to move left and right, was controlled with air pressure. For this study, the olive shakers were operated at about 450 KPa, which caused the heads with tines to move back and forth at ~1.2 Hz. Workers held the olive shaker with two hands with the hand at the opposite end from the shaker head controlling the on/off trigger. The long handle allowed workers to operate the shaker in the standing position with the handle oriented horizontally across at waist to chest height. It was also possible to raise the shaker head even above the worker's head to harvest blueberries at the top of 2-m-tall blueberry plants or to lean forward or bend the knees and point the shaker downward to harvest blueberries located close to the ground.

2.1.3. Over-the-Row Machine Harvesting (OTR)

Over-the-row blueberry harvesters are large machines designed to travel over rows of blueberry plants, one row at a time [11]. They typically have an inverted "U"-shaped frame tunnel large enough to move over large blueberry plants with sufficient clearance at the top. For this study, two harvesters (Oxbo 7440 and 8040, Oxbo International, Lynden, WA, USA) were used. Both harvesters detach blueberries with a rotary shaker mounted on each side of the tunnel. In both machines, detached berries are collected by catch plates on each side of the tunnel and roll onto a conveyor belt on each side of the harvester. Model 8040 is a single-drop harvester which transfers the fruit through a cleaning system to the back of machine, where it falls into a flat placed on a shelf. There are two workers on the platform on each side of the harvester. The first worker places an empty flat on the shelf underneath the end of the conveyor belt, waits for the flat to fill with blueberries, stops the belt by closing the hydraulic line, removes the filled flat from the shelf, and hands it to a second worker. The second worker takes the flat and stacks. Model 7440 is a top-load berry harvester designed to handle high fruit volume. The harvested fruit rolls onto a conveyor belt on each side of the harvester that moves the fruit to the back, where the fruit is transferred to an elevator bucket that moves the fruit to the platform at the top of the harvester. There the fruit is transferred to a horizontal conveyor belt, first going under a leaf/trash suction blower. The cleaned fruit is then transferred to another belt which moves the fruit horizontally. At the end of the conveyor belts, the fruit drops into a flat. Workers stationed on the top platform move empty flats that are stacked on pallets, and places them under the end of the conveyor belt to collect falling fruit in the flat. When the flat is full then the first worker lifts the filled flat and hands it to another worker. This worker then moves the filled flat to another pallet for stacking. There are shallow and deep flats used to collect machine-harvested blueberries. Shallow flats weigh about 4 kg when filled with blueberries. Deep flat weights 7 kg when filled with blueberries. For shallow trays, workers make six columns of flats on a pallet. As many as 12 flats are stacked, reaching a height of 1.4 m. When deep trays are used, workers also create six columns and the filled flats are stacked eight high, reaching a height of 1.8 m.

2.2. Ergonomics Intervention Techniques

2.2.1. Borg Category Ratio Scale (CR10)

The Borg category ratio scale (CR10) was used to determine workers' health and discomfort at work [16]. Borg CR10 is a subjective scale in which participants self-rate their level of physical pain or discomfort for various parts of the body on a scale from 0 to 10, where 0 represents no pain at all and

10 represents maximum pain. In the literature, numerous studies have validated the Borg scale with quantitative measurements of physiological responses (e.g., metabolic acidosis, ventilation, oxygen intake, heart rate, and respiration frequency) [17–19].

2.2.2. Rapid Upper Limb Assessment (RULA)

McAtamney and Corlett [20] developed the Rapid Upper Limb Assessment (RULA) to examine the level of risk of upper limb disorders for individual workers. RULA evaluates worker exposure to the position, force, and repetitive movement of different work postures that contribute to repetitive strain injuries (RSIs). It is a widely used measurement because it is easy and provides a quick assessment without requiring special equipment. RULA encompasses the postures of several body parts, including the wrists, arms, neck, shoulders, trunk, and legs, and accounts for force as well as repetition in those postures. The assessment consists of two sections, Section A and Section B. In Section A, scores are entered for the shoulders, arms, and wrists, while scores for the legs, neck, and trunk are entered in Section B. The posture scores from Sections A and B are then combined with the muscle use score and force score to obtain the grand score, which represents the level of MSD risk. The MSD risk score ranges from 1 to 7, where a score of 1 or 2 indicates an acceptable risk, 3 or 4 indicates low risk and may warrant further investigation, 5 or 6 indicates medium risk and requires investigation and change, and a score of 7 indicates high risk and requires immediate investigation and change.

2.2.3. Cumulative Trauma Disorders (CTD) Risk Assessment Model

The original version of the Cumulative Trauma Disorders (CTD) risk assessment model assesses the risk of CTDs in the upper extremities [21]. A simplified version of the CTD risk index appears in Niebel and Freivalds [22]. This version reduces the analysis complexity and time required for the assessment and is therefore more appropriate for evaluation in field study. This version assesses four factors: frequency, posture, force, and miscellaneous. The frequency factors index is calculated based on the number of hand motions per day scaled by the allowable limit of 10,000 daily hand motions. The posture factor is determined from grip types and the degree of deviation from the natural posture of the upper limbs. The force factor index is calculated based on percent maximum voluntary contraction (MVC) used in a given task and then scaled by 15 percent. Miscellaneous factors include the use of gloves, the presence of sharp edges on work contact surfaces, vibration exposure, and cold temperatures. These four factors are then weighted and summed to obtain a final CTD risk index as a job risk measure with a critical value of 1.

2.2.4. NIOSH Lifting Analysis

The National Institute for Occupational Safety and Health (NIOSH) published the Revised NIOSH lifting equation for evaluating the physical demands of two-handed manual lifting tasks based on biomechanical, psychophysical, physiological, and epidemiological factors in 1993 [23]. The NIOSH lifting equation has been used in several research studies to quantify biomechanical stressors from manual lifting and lowering tasks and has become the most commonly used job analysis method in the past two decades [24–27]. The lifting equation consists of two steps: (1) calculate the recommended weight limit (RWL, i.e., the maximum acceptable load), and (2) calculate the lifting index (*LI*, i.e., the relative estimate of the level of physical stress and MSD risk associated with lifting tasks) for a specified manual lifting task. The RWL is calculated depending on lifting conditions, e.g., hand location in relation to the body, vertical travel distance of hands, degree of symmetry in posture, frequency of lifting, work-rest duration pattern, and type of hand coupling:

$$RWL = 51 \times \left(\frac{10}{H}\right) \times (1 - 0.0075 \times |V - 30|) \times \left(0.82 \times \frac{1.8}{D}\right) \times FM \times 1 - 0.0035 \times A) \times CM \quad (1)$$

where the factors in the equation are:

H = horizontal load distance,
V = vertical load distance,
D = vertical displacement of the load,
FM = frequency multiplier,
A = asymmetric factor, and
CM = coupling multiplier.
LI is the ratio of the current load weight to the recommended load weight limit:

$$LI = \frac{Load\ Weight}{RWL} \tag{2}$$

An *LI* value of less than 1.0 indicates safe lifting without an increased risk of low back pain (LBP). An *LI* > 1 has been shown to be associated with LBP in previous studies [28,29]. Thus, the goal is to design all lifting jobs such that they result in an *LI* value of less than 1.0.

2.3. Video Recording Observation and Data Processing

Twelve farm workers (male: 11, female: 1) were observed during the harvesting process via video recording. For each worker, five-minute samples were recorded at randomly selected periods during the harvesting process. The video was captured at one-second intervals. About 3000 images per worker were captured. These images were categorized into groups for posture analysis. Figure 4 shows the example of categorized harvesting postures and the component ratio of each posture group. In Figure 4, the postures are defined as follows: for hand harvesting, high (H-A), middle (H-B), low (H-C), stretch (H-D), and squat (H-E); for semi-machine harvesting, push in high position (M-A), pull in high position (M-B), push in low position (M-C), pull in high position (M-D), and standing (M-E). EMG signals were obtained using disposable surface electrodes (Thought Technology TTL T3404; an active diameter of 1.0 cm and an inter-electrode distance of 2.4 cm). The electrodes were placed in the direction of the muscle fibers on the worker's skin after standard skin preparation. The EMG signals of each muscle were amplified and automatically converted into root mean square (RMS) values via a MyoScan-Pro sensor (Thought Technology Ltd, Montreal West, QC, Canada). The EMG signals for each participant's biceps brachialis (BB), anterior deltoid (AD), and L5 level on the trunk were normalized by:

$$EMG_{n,m}(\%MVC) = \frac{EMG_{task}}{EMG_{max}} \times 100 \tag{3}$$

where $EMG_{n,m}$ stands for the RMS EMG of each muscle for each participant and EMG_{max} is the maximum RMS EMG signal of each muscle obtained for all recorded postures for each participant. EMG_{task} represents the actual electromyographic activity of a specific muscle during the blueberry harvestings.

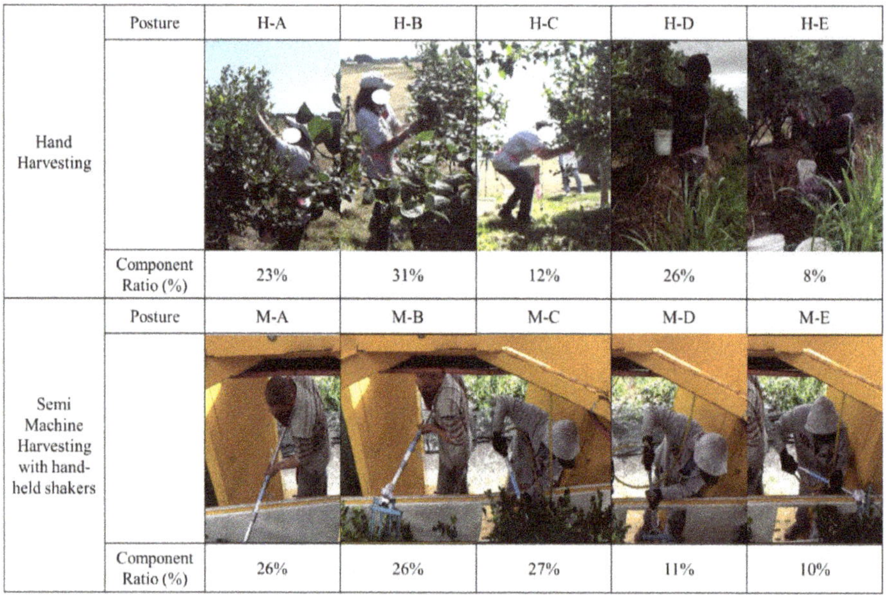

Figure 4. Examples of categorized harvesting postures and component ratios of each posture group: For hand harvesting, high (H-A), middle (H-B), low (H-C), stretch (H-D), and squat (H-E); for semi-machine harvesting, push in high position (M-A), pull in high position (M-B), push in low position (M-C), pull in high position (M-D), and standing (M-E).

3. Results

3.1. Demographic Result

Half of the subjects did not exercise regularly, and more than five had low education levels (i.e., through lower secondary school). Details of the blueberry harvester demographic data are shown in Table 1.

Table 1. Demographic information (SD = Standard deviation).

Characteristics			N (%)
Gender	Male		11
	Female		1
Age in years (mean ± SD)		31.3 ± 15.8	
Educational level	Primary school		5
	Lower secondary school		4
	Upper secondary school		3
Smoking Status	Yes		3
	No		9
Years smoking (mean ± SD)		3.0 ± 2.0	
Exercise frequency	None		6
	Some		3
	Often		3

On average, each blueberry farm laborer had worked 6.9 years at a rate of 8.3 h per day with a total of 17.9 min spent on breaks. Our analysis showed that half of the subjects experienced fatigue during daily work activities, and that most experienced the highest level of discomfort in the shoulder area. Table 2 shows the blueberry harvesters' physical ergonomic factors in their labor based on the questionnaire.

Table 2. Physical ergonomic factors of worker (n = 12, SD = Standard deviation).

Characteristics		N (%)
Working experience (mean ± SD) years		6.9 ± 7.5
Working days in a month (mean ± SD) days		26 ± 6.3
Working hours in a day (mean ± SD) hours		8.3 ± 1.6
Break duration (mean ± SD) minutes		17.9 ± 9.9
Experiencing work fatigue	Yes	6
	No	6
Frequency of work fatigue	Sometimes	6
	Frequently	5
	Always	1
Body parts with greatest discomfort	Shoulder	8
	Arm	4
	Hand	2
	Low back	6
	Leg	6
	Foot	6

3.2. Hand Harvesting

Regarding CTD risk assessment, both the mean of the frequency factor of 2.24 (SD: 0.12) and the posture factor of 1.43 (SD: 0.06) exceeded the safety threshold of 1.0, leading to a total risk value mean of 1.30 (SD: 0.05), which also exceeded 1.0. Table 3 shows the RULA scoring during blueberry harvesting. Score A represents wrist and arm scores and Score B refers to neck, trunk, and leg scores. Based on a Score A of 6 and a Score B of 7, posture H-A showed a RULA score of 7, the highest score among the evaluated postures, indicating that this harvesting posture needs further analysis and that immediate change should be implemented. Posture B displayed a RULA score of 3, the lowest RULA score among the evaluated postures, indicating that this harvesting posture needs to be changed or performed less frequently. Overall, the total component ratio of postures with a RULA score equal to or greater than 5 was 69%, indicating that more than half of the postures used during hand harvesting were medium risk and called for engineering and/or work method changes to reduce or eliminate MSD risk.

Table 3. Rapid Upper Limb Assessment (RULA) analysis results.

Posture	Component Ratio (%)	Score A	Score B	Grand Score
H-A	23	6	7	7
H-B	31	4	3	3
H-C	12	6	5	6
H-D	26	6	3	5
H-E	8	5	5	6

EMG was measured for five different positions including high (H-A), middle (H-B), low (H-C), lifting task, and normal standing posture. Overall, the low back and shoulders were commonly used for hand picking, while all muscles were used for lifting, as shown in Figure 5. The lifting task required the greatest muscle activity. For the harvesting task, the middle position required muscle activity like that of the standing position, resulting in similar EMG values between these two positions. The high position produced muscle activity three times higher than that of the standing posture. These results were consistent with the RULA postural analysis.

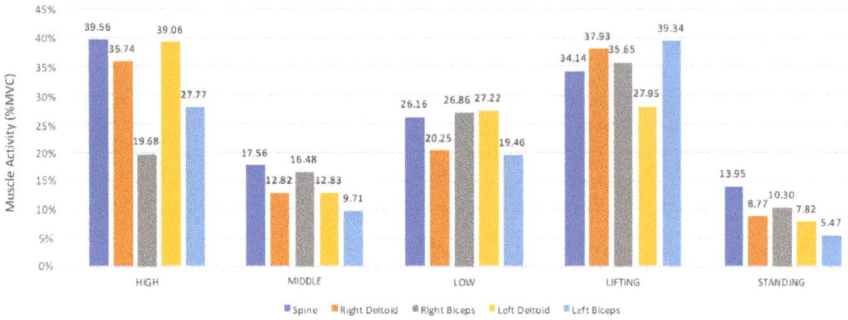

Figure 5. Muscle activity of five body regions by posture.

The mean Borg scale scores of the workers' perceived rate of discomfort is shown in Figure 6. The Borg scale scores showed a similar trend to that of the RULA analysis and EMG results. The low back exhibited the greatest mean Borg scale score (3.8), followed by the shoulder (3.5), neck (3.3), arm (3), and hand (2.7).

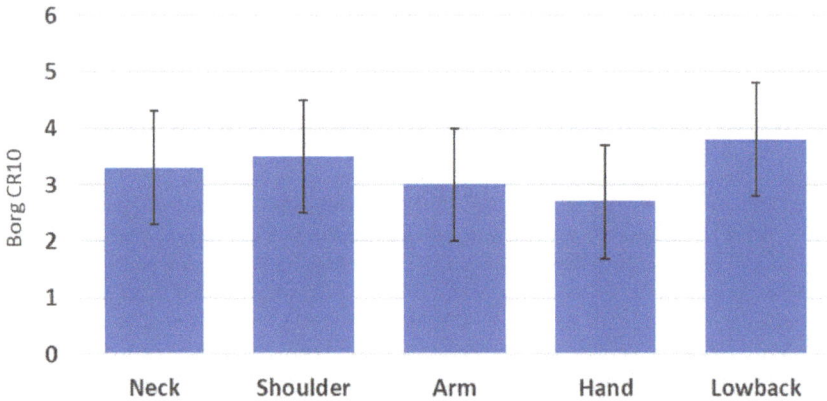

Figure 6. Subjective workload rating (Borg CR10).

Examples of the lifting postures are shown in Figure 7 and the results of the NIOSH lifting equations are shown in Table 4. The lifting task of picking up a bucket from the ground resulted in an *LI* value greater that in the NIOSH lifting equation, indicating that this lifting posture may increase the risk of developing lifting-related low back pain. This high *LI* value was due to the VM (Vertical Multiplier factor) since the worker needed to bend their trunk and knees to pick up the bucket from the ground without the aid of a raised pallet, which may contribute to low back pain.

Figure 7. Example of lifting posture during hand harvesting.

Table 4. NIOSH lifting equation results for lifting task show in Figure 7.

Results	Values
Recommend Weight Limit (RWL)	5.56 kg
Lifting Index (*LI*)	1.26

3.3. Semi-Machine Harvesting with Hand-Held Shakers

For semi-mechanical harvesting with hand-held power shakers, the result of CTD risk assessment for all postures was less than 1, indicating an acceptable risk of CTD for all tasks in the semi-machine harvesting process. Even though the posture factors exceeded the safety threshold of 1.0 with a mean value of 1.1 (SD: 0.14), the frequency factors were below the safety threshold of 1.0 with a mean value of 0.67 (SD: 0.07). Therefore, the total risk value was 0.73 (SD: 0.07). Table 5 shows the RULA scoring for semi-machine blueberry harvesting. Postures M-A and M-D each exhibited the highest possible RULA score of 7, mostly due to a side-bend of the trunk and an abducted shoulder posture. Since all postures showed a RULA score greater than 3, the machine design and/or worker posture should be adjusted to reduce or eliminate MSD risk.

Table 5. RULA analysis results.

Posture	Component Ratio (%)	Score A	Score B	Grand Score
M-A	26	5	6	7
M-B	26	4	4	4
M-C	27	6	4	6
M-D	11	6	6	7
M-E	10	4	4	4

For semi-machine harvesting with hand-held shakers, EMG values were measured for two positions: high (M-A and M-B) and low (M-C, M-D, and M-E). EMG results showed that both the high and low positions commonly required the right bicep to control the shaker, as shown in Figure 8. The low position required the left deltoid to push and pull the shaker, while the high position required the right deltoid to control the shaker and the left bicep to push and pull the shaker. Since the low posture closely resembled standing posture, low back EMG values showed little difference between high and low postures.

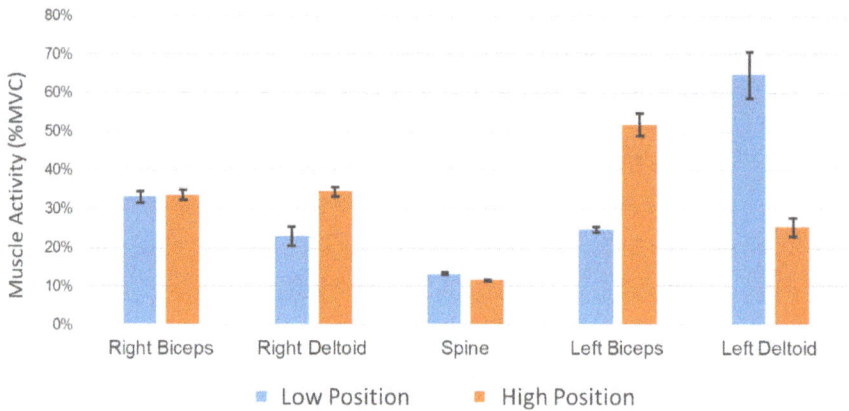

Figure 8. Muscle activity of five body regions by posture.

Example of the lifting posture are shown in Figure 9, with workers handling flats being filled with blueberries by an Oxbo 7240 harvester. The results of the NIOSH lifting equation are shown in Table 6. The lifting task at both origin and destination showed NIOSH equation values of less than 1, indicating a nominal risk to healthy employees.

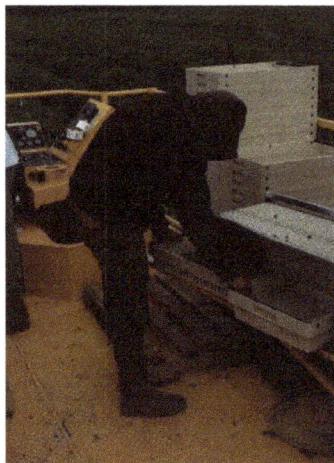

Figure 9. Example of lifting posture for handling flats filled with blueberries on an Oxbo 7240 blueberry harvester.

Table 6. NIOSH lifting equation results for lifting task.

Results	Values
Recommend Weight Limit (RWL)	11.88 kg
Lifting Index (*LI*)	0.80

3.4. OTR Machines

The lifting of full lugs weighing 8.2 kg (18 lb.) was studied in workers operating the Oxbo 7440 and the Oxbo 8040 machines. An example of the lifting posture is shown in Figure 10 and the results of the NIOSH lifting equations are shown in Table 7. The Oxbo 7440 could be unloaded either from the back of the conveyor system or from the side. The same was the case for the Oxbo 8040. In all cases the NIOSH lifting index (*LI*) was less than 1, and thus acceptable for an 8-hour day.

| (A) | (B) | (C) | (D) |

Figure 10. Example lifting postures: (**A**) Oxbo 7240 Back, (**B**) Oxbo 7440 Side, (**C**) Oxbo 8040 Side, and (**D**) Oxbo 8040 Back.

Table 7. National Institute for Occupational Safety and Health (NIOSH) lifting equation results for lifting task during over-the-row (OTR) harvesting.

Oxbo 7440 Back	Recommend Weight Limit (RWL)	9.52 kg
	Lifting Index (*LI*)	0.14
Oxbo 7440 Side	Recommend Weight Limit (RWL)	11.80 kg
	Lifting Index (*LI*)	0.69
Oxbo 8040 Side	Recommend Weight Limit (RWL)	9.12 kg
	Lifting Index (*LI*)	0.89
Oxbo 8040 Back	Recommend Weight Limit (RWL)	12.7 kg
	Lifting Index (*LI*)	0.64

EMG data is shown with respect to working conditions in Figure 11 and with respect to specific muscles in Figure 12. The Oxbo 8040 showed the worst EMG values with respect to working conditions, but the peaks (slightly over 40% MVC) were only for short periods of time. Greatest muscle loading was as expected for the arms and shoulders, but because peak values marginally exceeded 40% MVC and only for short periods of time, the loading posture associated with OTR machines was acceptable for an 8-hour day.

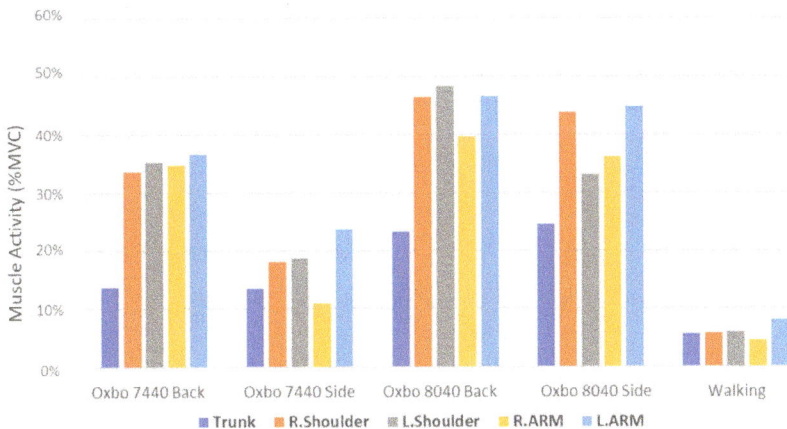

Figure 11. Muscle activity of five body regions by working condition (R.: Right; L.: Left).

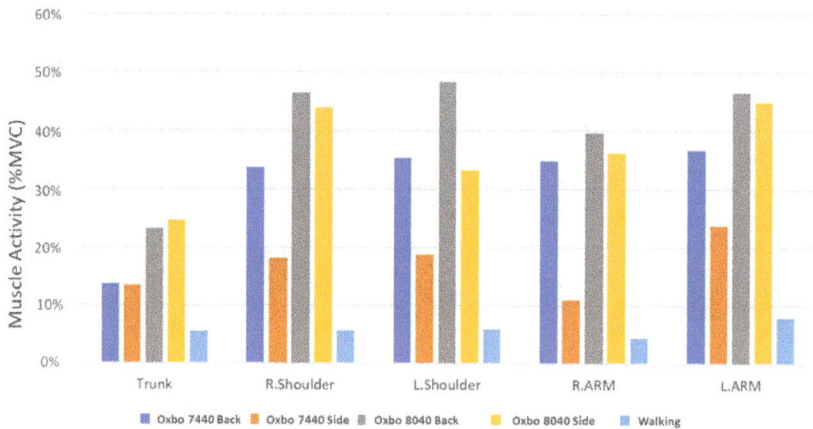

Figure 12. Muscle activity of five body regions by muscle (R.: Right; L.: Left).

4. Discussion

In this study, we conducted field research at several sites to investigate the risk factors of MSDs for each type of blueberry harvesting in terms of workers' postural loads and self-reported discomfort using ergonomics intervention techniques. Hand harvesting and semi-machine harvesting with hand-held shakers are difficult to analyze for fatigue, as there are many approaches with no clear limits. RULA values for hand picking ranged from 3 in the middle posture to 6 and 7 for low and high postures, respectively. Because the component ratio was over 50% in high and stretch position (H-A and H-D), we can conclude that this working posture contributes to workers' shoulder pain. The high scores of H-A and H-D suggest a need for redesign, but there is not much that can be done with hand harvesting. In addition, the RULA measurement assesses posture, but offers no clear insights on fatigue.

For semi-machine harvesting with hand-held shakers, the RULA numbers were also high, coming in between 4 and 7, but this was due to excessive repetition of the reaching motion. Once the standing positions on the vehicle and the length of shaker are adjusted, then working posture and RULA scores will improve. The EMG value of various muscles indicate the forces utilized. For hand picking, EMG values for the high posture slightly exceeded 40% MVC, which is higher than the recommended 15% for static contractions. The shakers required 30–50% MVC, but because this work is performed dynamically rather than in a static position, the recommended value of 15% MVC does not apply.

The best solution is to consider the Threshold Value Level [30] approach in conjunction with the Hand Activity Level (HAL). Two factors, force and HAL, determine whether the TLV is acceptable. The red region shown in Figure 13 represents hand harvesting, which has very high hand activity levels and lower EMG values but results in unacceptable fatigue values. The green region in Figure 13 represents the shakers, which have medium frequency and slightly higher EMG values but overall are still in the acceptable region for fatigue. The biggest problem with the shakers is the vibration, which can lead to fatigue and various risks for workers, especially in the upper limbs. Reducing the vibration with gel pads or gloves would decrease the risk to workers, as found in 2016 in the North Carolina studies that used simple foam as a stop-gap approach. However, according to a previous study [31], since anti-vibration gloves only extend the vibration exposure over time and therefore do not completely prevent the wearer from developing Hand-Arm Vibration Syndrome (HAVS), more research is required to conduct ergonomic interventions with reliable vibration data, not only with that obtained by subjective perceptions. In addition, according to a previous study conducted by Takeda et al. [11], hand-held shaker devices have not been widely adopted among blueberry growers due to harvest inefficiency and fatigue after prolonged use. Furthermore, Calvo et al. [32] pointed out that vibrating tools used in manual olive harvesting exceed the admitted limits of occupational repetitive action (OCRA) scores. Thus, workers who use vibrating tools in unnatural body postures are at risk for disorders of the upper limbs.

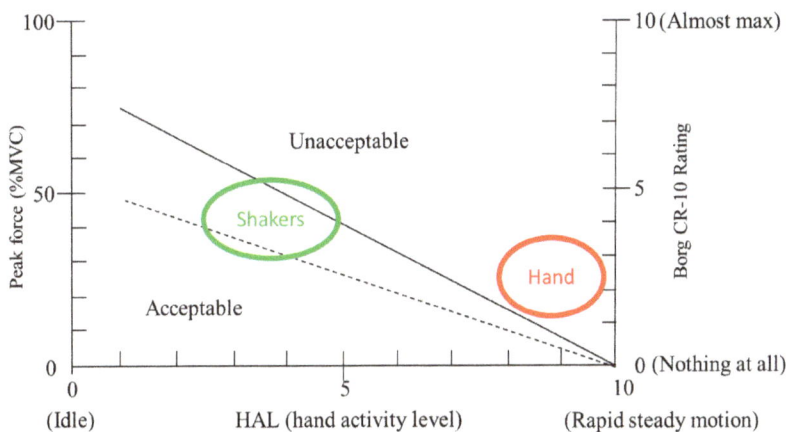

Figure 13. Threshold Level Values for Shakers and Hand Harvesting.

OTR harvesting eliminates repetitive hand motions and transfers all the work to simple lifting, for which there are clear approaches, such as the NIOSH lifting guidelines, to be used. In all cases, the lifting index was below 1 and therefore acceptable for an 8-hour day. This indicates that the automated harvester would be the best approach in minimizing worker loading and fatigue. However, it would be challenging for small- and medium-sized blueberry farms to purchase automated harvesters due to their high cost. Thus, collaborative efforts among health and safety professionals, engineers, social scientists, and ergonomists are needed to provide effective ergonomic interventions, including mechanical worker aid devices and tools as well as engineering and administrative controls such as programmed rest breaks, job rotation, and worker training.

Our study presents some features and limitations that should be noted. The first limitation of this study is the limited data to obtain reliable demographic results. Since most harvesters were migrants, the manager did not allow the collection of demographic data. The second limitation of this study is that it does not provide a solution that would eliminate risk factors of MSDs for hand and semi-machine harvesting. For example, this study does not suggest how to redesign the shaker to reduce vibration problems or how to train workers to adopt proper harvesting posture. However, the purpose of this study was to identify the risk factors of each type of blueberry harvesting. Thus, this limitation should not change the conclusions of this study. Further research will be required to find solutions for aid devices and tools as well as adequate training to reduce MSDs.

Author Contributions: E.K. wrote the manuscript. E.K. and A.F. collected and analyzed the data. F.T. and C.L designed and implemented the shakers and the mechanized harvester experiments. A.F. and F.T. edited the manuscript.

Funding: This research was funded by USDA National Institute of Food and Agriculture Specialty Crop Research Initiative, grant number 2014-51181-22383.

Conflicts of Interest: The authors declare no conflict of interest.

References

1. Bureau of Labor Statistics (BLS). *Occupational Injuries/Illnesses and Fatal Injuries Profiles*; Department of Labor, Bureau of Labor Statistics: Washington, DC, USA, 2016.
2. Putz-Anderson, V.; Bernard, B.P.; Burt, S.E.; Cole, L.L.; Fairfield-Estill, C.; Fine, L.J.; Grant, K.A.; Gjessing, C.; Jenkins, L.; Hurrell, J.J., Jr.; et al. *Musculoskeletal Disorders and Workplace Factors*; National Institute for Occupational Safety and Health (NIOSH): Cincinnati, OH, USA, 1997; p. 104.
3. Whelan, S.; Ruane, D.; McNamara, J.; Kinsella, A.; McNamara, A. Disability on Irish farms—A real concern. *J. Agromed.* **2009**, *14*, 157–163. [CrossRef] [PubMed]
4. McMillan, M.; Trask, C.; Dosman, J.; Hagel, L.; Pickett, W.; for the Saskatchewan Farm Injury Cohort Study Team. Prevalence of Musculoskeletal Disorders among Saskatchewan Farmers. *J. Agromed.* **2015**, *20*, 292–301. [CrossRef] [PubMed]
5. Walker-Bone, K.; Palmer, K.T. Musculoskeletal disorders in farmers and farm workers. *Occup. Med.* **2002**, *52*, 441–450. [CrossRef]
6. Fathallah, F.A. Musculoskeletal disorders in labor-intensive agriculture. *Appl. Ergon.* **2010**, *41*, 738–743. [CrossRef] [PubMed]
7. Choobineh, A.; Tabatabaee, S.H.; Behzadi, M. Musculoskeletal Problems among Workers of an Iranian Sugar-Producing Factory. *Int. J. Occup. Saf. Ergon.* **2009**, *15*, 419–424. [CrossRef] [PubMed]
8. Ng, Y.G.; Bahri, M.T.S.; Syah, M.Y.I.; Mori, I.; Hashim, Z. Ergonomics observation: Harvesting tasks at oil palm plantation. *J. Occup. Health* **2013**, *55*, 405–414. [CrossRef] [PubMed]
9. Kirkhorn, S.R.; Earle-Richardson, G.; Banks, R.J. Ergonomic risks and musculoskeletal disorders in production agriculture: Recommendations for effective research to practice. *J. Agromed.* **2010**, *15*, 281–299. [CrossRef] [PubMed]
10. Mora, D.C.; Miles, C.M.; Chen, H.; Quandt, S.A.; Summers, P.; Arcury, T.A. Prevalence of musculoskeletal disorders among immigrant Latino farmworkers and non-farmworkers in North Carolina. *Arch. Environ. Occup. Health* **2016**, *71*, 136–143. [CrossRef] [PubMed]
11. Takeda, F.; Yang, W.; Li, C.; Freivalds, A.; Sung, K.; Xu, R.; Hu, B.; Williamson, J.; Sargent, S. Applying New Technologies to Transform Blueberry Harvesting. *Agronomy* **2017**, *7*, 33. [CrossRef]
12. Brown, G.K.; Schulte, N.L.; Timm, E.J.; Beaudry, R.M.; Peterson, D.L.; Hancock, J.F.; Takeda, F. Estimates of mechanization effects on fresh blueberry quality. *Appl. Eng. Agric.* **1996**, *12*, 21–26. [CrossRef]
13. Mehra, L.K.; MacLean, D.D.; Savelle, A.T.; Scherm, H. Postharvest disease development on southern highbush blueberry fruit in relation to berry flesh type and harvest method. *Plant Dis.* **2013**, *97*, 213–221. [CrossRef]
14. Casamali, B.; Williamson, J.G.; Kovaleski, A.P.; Sargent, S.A.; Darnell, R.L. Mechanical Harvesting and Postharvest Storage of Two Southern Highbush Blueberry Cultivars Grafted onto Vaccinium arboreum Rootstocks. *HortScience* **2016**, *51*, 1503–1510. [CrossRef]
15. Lobos, G.A.; Moggia, C.; Retamales, J.B.; Sanchez, C. Effect of mechanized (self-propelled or shaker) vs. hand harvest on fruit quality of blueberries (*Vaccinium corymbosum* L.) in postharvest. In Proceedings of the X International Symposium on Vaccinium and Other Superfruits 1017, Maastricht, The Netherlands, 17–22 June 2012; pp. 141–145.
16. Borg, G. Psychophysical scaling with applications in physical work and the perception of exertion. *Scand. J. Work Environ. Health* **1990**, *16*, 55–58. [CrossRef] [PubMed]
17. Robertson, R.J.; Stanko, R.T.; Goss, F.L.; Spinal, R.J.; Reilly, J.J.; Greenawalt, K.D. Blood glucose extraction as a mediator of perceived exertion during prolonged exercise. *Eur. J. Appl. Physiol.* **1990**, *61*, 100–105. [CrossRef]
18. Gearhart, R.E.; Goss, F.L.; Lagally, K.M.; Jakicic, J.M.; Gallagher, J.; Robertson, R.J. Standardized scaling procedures for rating perceived exertion during resistance exercise. *J. Strength Cond. Res.* **2001**, *15*, 320–325. [PubMed]
19. Troiano, A.; Naddeo, F.; Sosso, E.; Camarota, G.; Merletti, R.; Mesin, L. Assessment of force and fatigue in isometric contractions of the upper trapezius muscle by surface EMG signal and perceived exertion scale. *Gait Posture* **2008**, *28*, 179–186. [CrossRef] [PubMed]
20. McAtamney, L.; Corlett, E.N. RULA: A survey method for the investigation of work-related upper limb disorders. *Appl. Ergon.* **1993**, *24*, 91–99. [CrossRef]

21. Seth, V.; Lee Weston, R.; Freivalds, A. Development of a cumulative trauma disorder risk assessment model for the upper extremities. *Int. J. Ind. Ergon.* **1999**, *23*, 281–291. [CrossRef]

22. Freivalds, A. *Niebel's Methods, Standards, and Work Design*; Mcgraw-Hill Higher Education: Boston, MA, USA, 2009; Volume 700.

23. Waters, T.R.; Putz-Anderson, V.; Garg, A.; Fine, L.J. Revised NIOSH equation for the design and evaluation of manual lifting tasks. *Ergonomics* **1993**, *36*, 749–776. [CrossRef] [PubMed]

24. Garg, A.; Boda, S. The NIOSH Lifting Equation and Low-Back Pain, Part 1. *Hum. Factors* **2014**, *56*, 6–28. [CrossRef] [PubMed]

25. Waters, T.R.; Putz-Anderson, V.; Baron, S. Methods for assessing the physical demands of manual lifting: A review and case study from warehousing. *Am. Ind. Hyg. Assoc. J.* **1998**, *59*, 871–881. [CrossRef] [PubMed]

26. van der Beek, A.J.; Mathiassen, S.E.; Windhorst, J.; Burdorf, A. An evaluation of methods assessing the physical demands of manual lifting in scaffolding. *Appl. Ergon.* **2005**, *36*, 213–222. [CrossRef] [PubMed]

27. Kucera, K.L.; Loomis, D.; Lipscomb, H.J.; Marshall, S.W.; Mirka, G.A.; Daniels, J.L. Ergonomic risk factors for low back pain in North Carolina crab pot and gill net commercial fishermen. *Am. J. Ind. Med.* **2009**, *52*, 311–321. [CrossRef] [PubMed]

28. Lavender, S.A.; Oleske, D.M.; Nicholson, L.; Andersson, G.B.; Hahn, J. Comparison of five methods used to determine low back disorder risk in a manufacturing environment. *Spine* **1999**, *24*, 1441. [CrossRef] [PubMed]

29. Waters, T.R.; Lu, M.-L.; Piacitelli, L.A.; Werren, D.; Deddens, J.A. Efficacy of the revised NIOSH lifting equation to predict risk of low back pain due to manual lifting: Expanded cross-sectional analysis. *J. Occup. Environ. Med.* **2011**, *53*, 1061–1067. [CrossRef] [PubMed]

30. ACGIH. *TLVs and BEIs Threshold Limit Values for Chemical Substances and Physical Agents and Biological Exposure Indices*; American Conference of Governmental Industrial Hygienists: Cincinnati, OH, USA, 2002.

31. Sampson, E.; Van Niekerk, J.L. Literature Survey on Anti-Vibration Gloves. 2003. Available online: https://scholar.google.com/scholar?hl=en&q=Literature+survey+on+anti-vibration+gloves&btnG=&as_sdt=1%2C39&as_sdtp= (accessed on 23 October 2018).

32. Calvo, A.; Romano, E.; Preti, C.; Schillaci, G.; Deboli, R. Upper limb disorders and hand-arm vibration risks with hand-held olive beaters. *Int. J. Ind. Ergon.* **2018**, *65*, 36–45. [CrossRef]

agronomy

MDPI

Article

Quality Parameter Levels of Strawberry Fruit in Response to Different Sound Waves at 1000 Hz with Different dB Values (95, 100, 105 dB)

Halil Ozkurt [1],* and Ozlem Altuntas [2]

[1]   Karaisali Vocational School, Cukurova University, Adana 01770, Turkey
[2]   Department of Horticulture, Inonu University, Malatya 44280, Turkey; ozlem.altuntas@inonu.edu.tr
*   Correspondence: ozkhalil@cu.edu.tr; Tel.: +90-0322-551-2057

Received: 8 May 2018; Accepted: 17 July 2018; Published: 23 July 2018

Abstract: All living organisms perceive mechanical signals, regardless of their taxonomic classifications or life habits. Because of their immobility, plants are influenced by a variety of environmental stresses, such as mechanical stress, during their growth and development. Plants develop physiological behaviors to adapt to their environment for long-term development and evolution. Sound-induced stress—an abiotic stress factor—is an example of mechanical stress and is caused by sound waves generated by different sources. This stress has a negative effect on the development and growth of plants. The strawberry plants evaluated in this study were exposed to three different sound intensity levels (95, 100, 105 dB) at a constant frequency of 1000 Hz. In strawberry plants, stress induced by sound waves is thought to trigger increased production of secondary metabolites as a defense mechanism. To determine the effect of sound applications, the fresh and dry weights of the roots and shoots were measured in strawberry plants, and the pH, total soluble solids (Brix), titratable acidity, vitamin C, total sugar, total acid, and total phenols were analyzed in the fruits. Results show that the sound stress, which was produced at a constant frequency (1000 Hz) and different sound levels (95, 100, 105 dB), affects the growth parameters of the plant and several quality parameters of the fruit.

Keywords: abiotic stress; fruit quality parameters; ascorbic acid; biomass; sound waves; frequency; dB

1. Introduction

Worldwide, strawberry (*Fragaria × ananassa*) is commonly consumed either in its fresh form or after it is processed. Strawberry reaches its full size and ripens within 30 days; it is a non-climacteric fruit. This growth period is dependent on light, temperature, soil composition, and some cultivation conditions [1]. In addition to being a fruit that is consumed for its taste, strawberry contains carbohydrates, vitamin C, and some antioxidant compounds (e.g., phenolics and flavonoids) [2]. Secondary plant metabolites are compounds with no fundamental roles in the life processes of plants, but they are important for the plant's ability to interact with its environment for adaptation and defense [3].

The ability to sense and respond to physical stimuli is of key importance to all living things. Light, temperature, and chemical signals are among the environmental stimuli detected by living organisms. Some of these stimuli are related to physical–mechanical stimuli (i.e., differences in mechanical forces or pressures detected by a living cell). Due to the force of gravity straining self-loading and inner growth, and mechanical loads of snow, ice, fruit, wind, rainfall, touch, sound, and hydration (turgor pressure) may be perceived by a cell. All living organisms perceive mechanical signals, regardless of their taxonomic classifications or life habits (sessile vs. motile). Because of their immobility, plants are influenced by environmental stress, such as mechanical stress, during their growth and development [4–7]. Plants develop physiological behaviors to adapt to the environment for long-term development and evolution. In previous studies, it was reported that plants are capable of responding to wind, touch, electric fields, magnetic fields, and ultraviolet rays [8,9]. Sato et al. [10] reported that, under mechanical stress, chloroplasts in plant cells were re-localized by an active motor system. Erner and Jaffe [11] reported that the contents of ethylene (C_2H_4) and abscisic acid (ABA) were increased in plants. They also reported that the gibberellin (GA) and indoleacetic acid (IAA) contents were decreased under mechanical stress. Secondary plant metabolites are compounds with no fundamental roles in the life processes of plants; however, they are vital elements in plants' interaction with the environment for adaptation and defense mechanisms.

Sound waves and sonication act as forms of abiotic stress on plants [12]. It is well known that plants absorb and resonate some sound frequencies from the external realm [13–15]. Sound waves had significant dual effects on the root development of *Actinidia chinensis* plantlets ($p < 0.05$). The root activity, total length, and the number of roots were increased by the stimulation from sound waves; however, the cell membrane permeability decreased Increasing ATP content in cells means that anabolism is strengthened. At 1 kHz and 100 dB, the soluble protein content and SOD activity were reported to increase. On the other hand, when sound wave stimulation exceeded 1 kHz and 100 dB, these indices were reduced [16–18].

2. Sound Waves and Sound Magnitude

Sound is a mechanical vibration wave that travels in a medium that consists of certain materials. According to physicists, 'sound' is the molecular diffusion of an energy source in air medium. Sound consists of vibrations in the air that are sensed by our brains after traveling as waves in a medium and stimulating our ears. Sound waves take the form of sinus waves. The distance between two peaks is the wavelength, and the number of wave peaks measured within 1 s is the frequency (Figure 1a,b). In other words, the frequency of a wave depends on the frequency of the vibrations of the particles in the medium (e.g., air) through which the wave travels. Frequency is computed by measuring the vibrations in time. The number of vibrations in 1 s is expressed in units of Hertz (1 Hertz = 1 cycle/s).

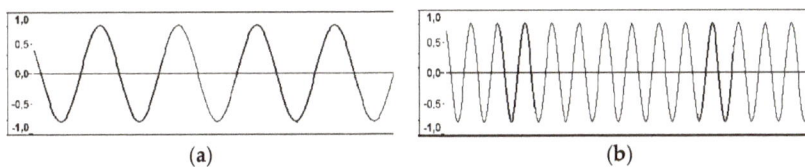

Figure 1. (a) Low-frequency sinus wave; (b) high-frequency sinus wave.

The term 'sound magnitude level' refers to the logarithm of an energy-physical magnitude. A decibel measures the perceived sound level; it is the noise level unit. The basic sound magnitude parameter for the sound magnitude level is I_0, which is the hearing limit at 1000 Hz:

$$L_I = 10 \cdot \text{Log} \frac{I}{I_0} \cdot (\text{dB}) \tag{1}$$

The magnitude of sound is in proportion to the square of the sound pressure ($I \sim p^2$), and the level of the sound pressure is found with the equation:

$$L_p = 10 \cdot \log \frac{p^2}{p_0^2} = 20 \cdot \log \frac{p}{p_0} \cdot (\text{dB}) \tag{2}$$

Here, the basic sound pressure p_0 at 1000 Hz is accepted as 2×10^{-5} Pa at the hearing limit. Sound measurement devices directly show the sound pressure level in decibel units using these equations. The sound power level, on the other hand, is the measurement of the sound power diffusing in any direction from the source and is expressed in a logarithmic manner, as in the case of sound magnitude level.

Using a reference power of 1 picowatt/m², i.e., 10^{-12} W/m², the SPL (sound power level) is calculated as:

$$SPL = 10 \cdot \log(\frac{W_{Source}}{W_{Reference}}) \tag{3}$$

W_{Source} = The total power diffused by the source
$W_{Reference} = 10^{-12}$ W/m² [19].

Strawberry is a fruit that is consumed worldwide and is cultivated in open and greenhouse systems. It was determined in previous studies conducted on a variety of fruits and vegetables that stress conditions cause increased production of secondary metabolites as a defense mechanism. However, since a plant's development is negatively affected when it is exposed to abiotic and biotic stresses, these stresses cause losses in yield. To date, no studies have investigated the effect of sound waves on the quality parameters of the strawberry fruit. Under stress conditions, plants increase production of compounds such as phenolic compounds and ascorbic acid to protect themselves. Therefore, the concentrations of nutritional compounds (which accumulate because of sound stress) are expected to increase under stress. In this study, we aimed to increase the quality parameters in the strawberry via sound waves at 1000 Hz and three different frequencies (95, 100, and 105 dB) without harming the plant. The effects of sound stress on the strawberry fruit's total soluble solids, titratable acidity, total sugars, total acids, pH total phenolic, and ascorbic acid were measured.

3. Materials and Methods

In the present study, a sound amplifier that was capable of transmitting sounds at different decibel values, along with a decibel indicator, was used. A signal generator that was capable of being adjusted was used as a frequency oscillator for creating the 1000 Hz frequency. Three 2×2 m chambers that were prepared specifically for the experiment and whose four sides could be opened were used. Furthermore, a sound level meter (noise measurement device) was used as a sound measurement device. Also, speakers that could produce 360° sound were used. The glass was 4 mm thick in the chambers. Between the glass was a 10.5 mm space.

Strawberry plants (*Fragaria × ananassa* Duch. cv Festival) in 4 L pots of turf + perlite (1:1) were used as the plant material. Hoagland nutrient solution ((M): Ca(NO$_3$)$_2$·4H$_2$O, 3.0×10^{-3}; K$_2$SO$_4$, 0.90×10^{-3}; MgSO$_4$·7H$_2$O, 1.0×10^{-3}; KH$_2$PO$_4$, 0.2×10^{-3}; H$_3$BO$_3$, 1.0×10^{-5}; 10^{-4} M FeEDTA, MnSO$_4$·H$_2$O, 1.0×10^{-6}; CuSO$_4$·5H$_2$O, 1.0×10^{-7}; (NH)$_6$Mo$_7$O$_{24}$·4H$_2$O, 1.0×10^{-8}; ZnSO$_4$·7H$_2$O, 1×10^{-6}) was used to water the plants [20]. The plants were placed in the sound chambers for the purpose of measuring the effect of sound stress.

A total of four special sound chambers were used to conduct the study. The chambers were placed in a plastic greenhouse at Research Fields of Cukurova University, Karaisalı Vocational High School (36°59′ N, 35°18′ E, 20 m above sea level), Adana, Turkey. A randomized complete block experimental design was applied in the study (3 replicates, 10 plants in each replicate). Then, different sound waves were directed at the pots. The speaker was 65 cm from each pot (Figure 2). Sound magnitudes of 95, 100, and 105 dB were directed at the pots in the sound chambers. For 30 days, the sound waves were delivered once per day for 1 h in the morning, between 10:00 and 11:00 a.m. The plants were in closed chambers when the sound waves were emitted.

It has been stated by specialists that being exposed to sound exceeding 85 dB might be dangerous. For this reason, dB values above 85 dB were selected. To compare results, control plants were placed in a chamber in which no sound applications were made.

The experiment was designed as a randomized complete block experimental design with 3 replicates, 10 plants in each replicate. A total of four chambers were used. The chambers could only detect the sound within, as the chambers were built in such a way that no other sound could enter. IBM SPSS Statistics 20 software was used for data analysis. The mean values of the fruit parameters for the three sound frequencies were compared using an ANOVA test. The effects of sound on the fruit parameters were considered significant at $p \leq 0.05$.

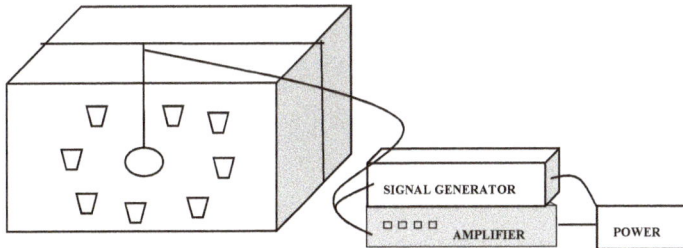

Figure 2. Illustration of the trial design.

3.1. Measurements and Analyses in Strawberry Plants and Fruits

The trial was started on 16 April 2014 and finalized on 16 May 2014. The fresh and dry weights of the roots in strawberry plants and the fresh and dry weights of the green parts were measured every 10 days for a total of three measurement points throughout the trial. The pH, total soluble solids (Brix), titratable acidity, vitamin C, total sugar, total acid, and total phenol contents of the fruits were analyzed. For the purpose of preparing ultrapure water (18.2 MΩ cm), the Millipore System (Millipore Corp., Bedford, MA, USA) was used. The chromatography reagent standards and solvents were obtained from Sigma Chemical, Co. (St. Louis, MO, USA).

3.2. Determination of Total Soluble Solids (TSS) and Titratable Acidity (TTA)

A hand-type refractometer (ATAGO ATC-1, Tokyo, Japan) was used to determine the total soluble solids in the juice of each sample. For the purpose of determining the total titratable acidity levels, the acid–base titration method was applied. The juice (1 mL) and distilled water (50 mL) were added to a conical bottle to titrate with aqueous NaOH (0.1 N) to obtain pH 8.1. Total acid content was determined in citric acid equivalents and is reported as the mean value of triplicate analyses.

3.3. Extraction of Sugars and Acids (TS and TA)

One gram of the sample was weighed and powdered with liquid nitrogen. The sample was added to 20 mL of aqueous ethanol (80%, *v/v*) and the solution placed in a screw-cap Eppendorf tube and then in an ultrasonic bath where it was sonicated for 15 min at 80 °C. It was then filtered through filter paper (the extraction was repeated three more times). The filtered extracts were mixed and evaporated in a boiling water bath until dry. Distilled water (2 mL) was used to dissolve the precipitation, and the resulting solution was filtered using Whatman nylon syringe filters (0.45 μm pore size, 13 mm diameter) before HPLC analysis. For organic acid extraction, liquid nitrogen was used to powder the homogenate (1 g of frozen sample), which was then weighed. Then, it was mixed with 20 mL aqueous metaphosphoric acid (3%) at room temperature for 30 min with a shaker. The mixture was then filtered and its volume was increased to 25 mL using the same solvent. It was then used for HPLC analysis [21].

3.4. HPLC of Organic Acid and Sugars

There is a built-in degasser, pump, and controller coupled to a photodiode array detector (Shimadzu SPD 10A *vp*) in the high-performance liquid chromatographic apparatus (Shimadzu LC 10A *vp*, Kyoto, Japan). The device also has an automatic injector with a 20 µL injection volume and is interfaced with a computer with Class VP Chromatography Manager Software (Shimadzu, Japan). The separation process was performed with a 250 × 4.6 mm i.d., 5 µm, reverse-phase Ultrasphere ODS analytical column (Beckman, Fullerton, CA, USA). The column was run at room temperature with a flow rate of 1 mL min^{-1}. The process was performed with a 0.1 a.u.f.s. sensitivity (wavelengths between 200 and 360 nm). The elution was isocratic with 0.5% aqueous metaphosphoric acid. The retention times of the components were compared using an in-house PDA library to identify the components that had authentic standards under analytical conditions and UV spectra. Between injections, there was a 10 min equilibrium time. The Shimadzu LC-10 A *vp* device was used for separating the sugar on a 150 × 4.6 mm i.d., 5 µm, reverse-phase Nucleosil NH$_2$ analytical column (Shimadzu, Tokyo, Japan) at room temperature with a 1 mL min^{-1} flow rate [21].

3.5. Determination of Total Phenolic Content (TPC) and Ascorbic Acid (AA) in Strawberry Fruits

In order to determine the total phenolic content (TPC), the Folin–Ciocalteu method was employed. After homogenization with a T18, IKA Homogenizer, Germany, 5 g of the frozen fruits with 25 mL ethanol was centrifuged at 3500× *g* for 3 min. Filter paper was used to filter the supernatant. Then, 2 mL of 10% Folin–Ciocalteu reagent was added to 0.4 mL of the extract. After this, it was left idle for 2–3 min. Finally, 1.6 mL (7.5%) of Na$_2$CO$_3$ solution was added to the mixture, which was incubated for 1 h in the dark, after which is was measured at 765 nm in a spectrophotometer (UV-1201, Shimadzu, Kyoto, Japan) against a blank solution (0.4 mL water + 2 mL Folin–Ciocalteu reagent + 1.6 mL Na$_2$CO$_3$). Using the gallic acid standard, the total phenolic content was computed as 1 mg gallic acid equivalent (GAE) 100 g^{-1}. The results are reported as mg/gallic acid equivalents per gram/dry weight. The Merck RQflex reflectometer was employed to analyze the ascorbic acid content (AA) in the samples by adopting the protocol for the juice of red fruit. The results are given as mg ascorbic acid/100 g fresh sample [21].

4. Results and Discussion

4.1. Results of the Weight Measurements in Strawberry Plants in Fresh and Dry Roots and Shoots

Table 1 shows that the difference between the fresh and dry weight values of the roots was statistically significant for all three measurement dates. As the sound level increased, the root growth regressed. For the measurement that was made 10 days after the initial sound application, the lowest root weight was determined to belong to plants subjected to 105 dB sound, followed by those exposed to the 100 dB level. The highest fresh and dry root weight values were found to be in the control plants. The fresh and dry root weights of plants at the 95 dB level were equal to those of the control plants. For the root weight measurements that were made 20 days after the initial sound application, it was determined that the lowest fresh and dry values were in the plants that received sound at the 105 dB level, and the highest values were measured in the control plants. With respect to fresh weight measurements, the sound levels fall into different statistical groups; for dry weight measurements, only the application at 105 dB is significantly different from the others. Similar results were obtained for the measurements that were made 30 days after the initial sound application. For the fresh root weight, the control plants and the plants at the 95 dB level had similar values and fall into the same statistical group. The weight decreased at 100 dB, the control group plants are included in the group with the plants at 95 dB; the treatment at 105 dB, having the lowest root weight values, is in its own group as it was significantly different from all other applications (Table 1).

Table 1. Fresh and dry weight changes in the roots and shoots. Measurements were made at 10-day intervals in plants that were exposed to different sound levels.

Sound Intensity Treatments	10 Days after the Sound Treatment				20 Days after the Sound Treatment				30 Days after the Sound Treatment			
	RFW (g)	RDW (g)	SFW (g)	SDW (g)	RFW (g)	RDW (g)	SFW (g)	SDW (g)	RFW (g)	RDW (g)	SFW (g)	SDW (g)
95 dB	7.94 a	1.10 a	10.92	2.61	9.50 ab	1.38 a	17.84 ab	3.10 a	12.50 a	1.72 a	21.25 a	3.64
100 dB	5.77 b	0.85 b	10.33	2.56	8.98 b	1.32 a	16.32 b	3.23 a	9.32 b	1.40 ab	19.18 b	3.37
105 dB	4.74 b	0.64 b	9.13	2.20	6.30 c	0.78 b	15.05 b	2.95 b	8.86 c	0.96 b	18.52 b	3.22
Control	7.34 a	1.15 a	11.15	2.78	10.57 a	1.43 a	19.52 a	3.40 a	13.03 a	1.77 a	22.92 a	3.59

The means in the columns followed by different letters are significantly different ($p < 0.05$); RFW: root fresh weight; RDW: root dry weight; SFW: shoot fresh weight; SDW: shoot dry weight.

The measurement results in Table 1 for fresh and dry weights of shoots show there was no statistically significant difference in these values between the applications for the measurements made 10 days after the onset of the sound application. However, like the root weights, the sound magnitude negatively affected shoot weight, the control plants had the highest values, and weight decreased at 105 dB. In the second measurement, made 20 days after the initial sound application, the difference between the applications was found to be statistically significant in terms of shoot fresh and dry weight values. For the root fresh weight results, while 100 and 105 dB treatments are in the same statistical group with the lowest values, the control plants, with the highest values, falls into a separate group, showing a significant difference compared to the 100 and 105 dB level applications. The 95 dB sound application is in an intermediary group. For the measurements made 30 days after the initial sound application, the fresh shoot weight results were determined to be significant at a statistical level. The control plants had the highest fresh weight values, and the plants at 95 dB had the second highest. The lowest shoot fresh weights were found in plants at the 100 dB and 105 dB levels; they were significantly different from the other two applications (control and 95 dB). For the last measurement date, the differences between the applications were not found to be significant at a statistical level in terms of shoot dry weight values. However, it was determined that the shoot dry weight values were lower in the plants that were exposed to sounds at high levels (Table 1 and Figure 3).

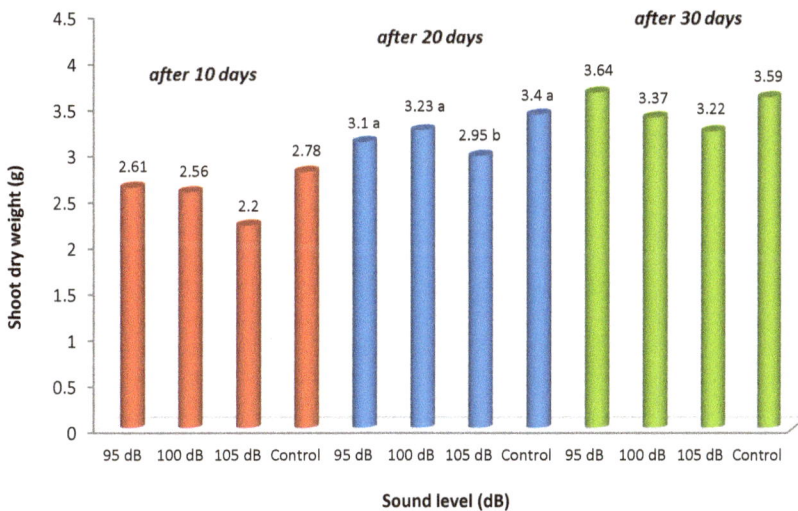

Figure 3. Shoot dry weight changes. Measurements were made at 10-day intervals in plants that were exposed to different sound levels. The means of different letters are significantly different ($p < 0.05$).

At 105 dB, as the sound level increased, the rate of the decrease in root dry weights was 44–46% for all measurement dates compared with the control plants. The decrease in root dry weight values at the 100 dB sound level was determined to occur at a rate of 8–26% compared to the control plants. The shoot dry weight values at 105 dB decreased at a rate of 10.3–21% compared to the control plants. The shoot dry weight values at the 100 dB sound level decreased at a rate of 5–8% compared to the control plants (Figure 3). The 95 dB level is included in the same group as the controls for most measurements, although 95 dB did not cause much weight loss in the roots and shoots.

Energy metabolism (for example, sugar, lipid, and photosynthesis) is influenced as abiotic stress increases [22–25]. For this reason, it is possible that metabolic responses to abiotic stress are gradual and complex. Abiotic stress also influences various cellular processes like growth, photosynthesis, carbon partitioning, carbohydrate–lipid metabolism, osmotic homeostasis, protein synthesis, and gene expression [26–28].

On the other hand, DNA damage occurs due to UVBR, and photosynthesis, secondary metabolites, and the synthesis of phenolic compounds are reduced [29–31].

4.2. Results of the Analysis of Strawberry Plants

When we consider the results of the analysis made 30 days after the initial sound applications at different levels (in Table 2), we see that pH values were similar for all applications, ranging between 3.36 and 3.39. Similarly, total soluble solids (TSS) and titratable acidity (TTA) values were not significantly affected by the sound applications at different levels. The TTS results were similar, ranging between 8.2 and 8.4. The TTA results were between 7.35 and 7.42. Although there were differences in total acid (TA) for different applications, this difference was not at a statistically significant level. As the sound level increased, the acid rates in the fruits increased. While the total acid was 21.36 in the fruits at the 105 dB sound level, this value was 19.55 in the control plants (Table 2).

Table 2. Measurements of several quality parameters of the fruits taken from plants 30 days after exposure to different sound levels

Sound Intensity Treatments	pH	TSS (%)	TTA (g kg^{-1})	TS (g kg^{-1})	TA (g kg^{-1})	TPC (mg 100 g^{-1} Gallic Acid)	AA (mg 100 g^{-1})
95 dB	3.36	8.4	7.37	59.33 ab	19.82	279.5 b	28.3 b
100 dB	3.38	8.3	7.41	62.06 a	20.62	282.0 ab	31.8 a
105 dB	3.39	8.4	7.42	63.25 a	21.36	288.0 a	32.7 a
Control	3.36	8.2	7.35	57.95 b	19.55	275.5 b	27.4 b

The means in the columns followed by different letters are significantly different ($p < 0.05$); TSS: total soluble solids; TS: total sugars; TTA: titratable acidity; TA: total acids; TPC: total phenolic content; AA: ascorbic acid.

It was determined that different sound levels caused statistically significant differences in total sugar (TS), total phenol content (TPC), and ascorbic acid (AA). Total sugar increased for the applications of 105 and 100 dB sound levels, falling into the same statistical group with values of 62.06 and 63.25. The control group fruits, falling into a separate group, were determined to have 57.95 total sugar. Plants exposed to the 95 dB sound level had total sugar that falls into the intermediary group with a value of 59.33. Total phenol content also increased with the sound magnitude; the phenol contents of the fruits at 105 dB was determined to be 288.0 and is included in a separate group. Total phenol content was 275.5 in the control group plants, which was the lowest value, and it was 279.5 for plants at the 95 dB sound level; these values are in the same statistical group. Total phenol content was 282.0 for the application at 100 dB, which is in the intermediary group. Ascorbic acid (AA) results were similar to the patterns for TPC. While the ascorbic acid of the control plants was 27.4, which is the lowest value, it was 28.3 at 95 dB; the control and 95 dB treatment are in the same group. The ascorbic acid values for the 100 dB and 105 dB sound levels were 31.8 and 32.7, respectively. These are the highest values, and these two applications were found to be in the same group (Table 2).

Total sugar (TS), total phenol content (TPC), and ascorbic acid (AA), which were detected at statistically significant levels in the fruit analyses, increased with the increasing sound magnitudes. Total sugar increased at a rate of 9% in strawberry fruits at 105 dB compared to the control group, and it increased at a rate of 7% at 100 dB (Figure 4). Total phenol content increased at a rate of 4.5% at 105 dB and at a rate of 2.4% at 100 dB. Ascorbic acid increased at a rate of 19% at 105 dB compared to the control group and at a rate of 16% at 100 dB (Table 2 and Figure 5).

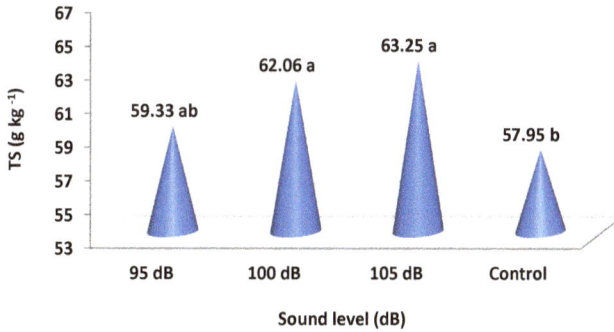

Figure 4. Total sugars (TS) in strawberry fruits 30 days after exposure to different sound waves. The means of different letters are significantly different ($p < 0.05$).

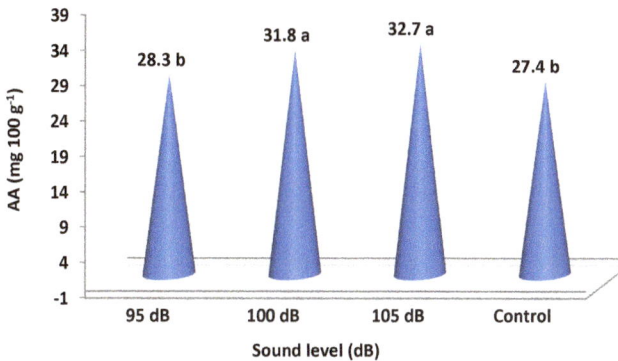

Figure 5. Total ascorbic acid (AA) in strawberry fruits 30 days after exposure to different sound waves. The means of different letters are significantly different ($p < 0.05$).

The diversity in the structure and function of secondary metabolites makes them necessary because they are of critical importance to the survival of plants under stress conditions [32]. Many environmental stresses (high/low temperature, drought, alkalinity, salinity, UV stress, and pathogen infection) have the potential for damaging plants [3]. In laboratory conditions, production of secondary metabolites was induced anew by using elicitation [3,33]. Several researchers applied various elicitors to improve secondary metabolite production in cultures of plant cells, tissues, and organs [34]. Nutrient stress has an important influence on phenolic levels in plant tissues [3,34]. Pathogen attack, UV irradiation, high-intensity light, wounds, nutrient deficiency, temperature, herbicide treatment, and other environmental stress factors increase the accumulation of phenylpropanoid [35–37]. The effects of some secondary plant products on growing conditions are high in terms of the metabolic pathways that are responsible for accumulation of the related natural products.

Sugar accumulation is a common result of abiotic stress (e.g., glucose, fructose, and sucrose accumulate, along with other osmolytes, during cold treatment [38–40]). It was traditionally believed that osmolyte accumulation protected plant cells (either by osmotic adjustment or by stabilizing membranes and proteins); however, in time, another role of osmolytes was proposed to be the regulation of redox or sugar signaling. These influences might, for instance, involve hexokinase-dependent signaling or interactions between trehalose synthesis and sugar and ABA signaling [41,42]. It was been reported that ABA synthesis and signaling are important components in sugar signaling. In plants, sugars play important roles as both nutrients and signal molecules. Both glucose and sucrose are recognized as pivotal integrating regulatory molecules that control gene expression related to plant metabolism, stress resistance, growth, and development [43–45]. It was recently proposed that soluble sugars, especially when they are present at higher concentrations, might act as reactive oxygen species (ROS) scavengers themselves [46]. All abiotic stresses generate ROS, potentially leading to oxidative damage affecting crop yield and quality. In addition to the well-known classical antioxidant mechanisms, sugars and sugar-metabolizing enzymes have entered the picture as important players in the defense against oxidative stress [47].

Abiotic stress may be used in preharvest activities to improve the quality and yield of products [48]. For instance, vitamin C may be improved in plants that are exposed to high-intensity light or in plants that have less frequent irrigation [49].

The environment affects the ascorbic acid concentration in the fruits and leaves [50,51]. The level of synthesis controls the regulation of ascorbate levels in cells [52,53]. Recycling and degradation (Pallanca and Smirnoff, Green and Fry [53,54]) and transport of this molecule in cells or between organs (Horemans, Foyer, and Asard [55]) are also controlled by the synthesis. During the stress response and the adaptation to stress, the recycling pathway is important. Reduced ascorbate, an antioxidant, is oxidized into an unstable radical (monodehydroascorbate) in oxidative stress conditions; the oxidized molecules then dissociate into ascorbate and dehydroascorbate.

Alessandra Ferrandino and Claudio Lovisolo [56] claimed that abiotic stress modified the growth and development in all plant organs of grapevine plants. At the berry level, the response to abiotic stress drives the accumulation of secondary metabolites in berry pulps, seeds, and skins as a defense against cell damage. Viticultural trials may be designed to control plant stress response to increase secondary metabolite concentrations.

Exposure of plants to unfavorable environmental conditions (e.g., heavy metals, drought, nutrient deficiency, salt stress) can increase the production of reactive oxygen species (ROS). To protect themselves against these toxic oxygen intermediates, plants employ antioxidant defense systems [57]. To control the levels of ROS and to protect cells under stress conditions, plant tissues contain several enzymes that scavenge ROS (SOD, CAT, peroxidases, and glutathione peroxidase), detoxify LP products (glutathione-*S*-transferases, phospholipid-hydroperoxide glutathione peroxidase, and ascorbate peroxidase), and a network of low molecular mass antioxidants (ascorbate, glutathione, phenolic compounds, and tocopherols) [57]. Secondary metabolites are involved in protective functions in response to both biotic and abiotic stress conditions. As determined in a recent study, environmental factors increase the concentrations of phytochemicals [58]. Stress predominates among all the factors that enhance the concentrations of phytochemicals in fruits and vegetables. This makes sense when it is considered that all stress types (biotic/abiotic) are conducive to oxidative stress in plants [59], and oxidative signaling controls synthesis and accumulation of secondary metabolites [60]. Plants produce phenolic compounds as a defensive mechanism to biotic/abiotic stresses [61].

In the strawberry plant in the fruit development stage, sound frequency stress promoted metabolite accumulation, which resulted in an improvement in fruit quality.

5. Conclusions

In the present study, different sound waves with different decibel values at a constant frequency were applied, and they caused decreases in the weight of roots and the green parts of the strawberry plants. However, these different sound waves also caused some increases in several quality parameters in the fruits at a statistically significant level. This increase, which also enhances the value of the fruit and is important for human health, is positive. Salinity, drought, high temperature, and irreversible abiotic stress factors reduce the growth and development of the plants, causing major losses in terms of yield; plants even die under continuous stress. In further studies, greenhouse trials may be performed at sound levels that do not affect plant development and yield at significant levels but increase fruit quality. In this way, sound applications that do not damage the soil and plant but increase the quality parameters in a positive manner may be recommended for greenhouse cultivation. In the present study, the weight losses were found to be greater in plants at the 105 dB sound level. At 95 dB, on the other hand, the results were close to those of the control group plants which were not exposed to sound. For this reason, 1000 Hz and 100 dB sound levels may be used for the strawberry plant to increase the quality of its fruits. However, in the future, it should be investigated whether it affects the yield in greenhouse designs.

Author Contributions: H.O. and O.A. conceived and designed the experiments; H.O. performed parts of physics and sound system of the experiments; O.A. performed the parts related to plants and O.A. analyzed the data; O.A. contributed reagents/materials/analysis tools; H.O. and O.A. wrote the paper.

Funding: We thank Cukurova University Scientific Research Projects Directorate for their financial support. And also, we thank to Ebru Kafkas and her laboratory team for strawberry fruits analyses.

Conflicts of Interest: The authors declare no conflict of interest.

References

1. Klein, B.P.; Perry, A.K. Ascorbic acid and vitamin A activity in selected vegetables from different geographical areas of the United States. *J. Food Sci.* **1982**, *47*, 941–945. [CrossRef]
2. Robards, K.; Prenzler, P.D.; Tucker, G.; Swatsitang, P.; Glover, W. Phenolic compounds and their role in oxidative processes in fruits. *Food Chem.* **1999**, *66*, 401–436. [CrossRef]
3. Akula, R.; Ravishankar, G.A. Influence of abiotic stress signals on secondary metabolites in plants. *Plant Signal. Behav.* **2011**, *6*, 1720–1731. [CrossRef] [PubMed]
4. Ozkurt, H.; Altuntas, O. The Effect of Sound Waves at Different Frequencies upon the Plant Element Nutritional Uptake of Snake Plant (*Sansevieria trifasciata*) Plants. *Indian J. Sci. Technol.* **2016**, *9*, 48–55. [CrossRef]
5. Ozkurt, H.; Altuntas, O.; Bozdogan, E. The Effects of Sound Waves upon Plant Nutrient Elements Uptake of Sword Fern (*Nephrolepis exaltata*) Plants. *J. Basic Appl. Sci. Res.* **2016**, *6*, 9–15.
6. Jaffe, M.J.; Leopold, A.C.; Staples, R.C. Thigmo responses inplants and fungi. *Am. J. Bot.* **2002**, *89*, 375–382. [CrossRef] [PubMed]
7. Baluška, F.; Šamaj, J.; Wojtazek, P.; Volkmann, D.; Menzel, D. Cytoskeleton–plasma membrane–cell wall continuum in plants. Emerging links revisited. *Plant Physiol.* **2003**, *133*, 483–491. [CrossRef] [PubMed]
8. Mary, M.; Braam, J. The Arabidopsis TCH4 xyloglucan endotransglyco-sylase. *Plant Physiol.* **1997**, *115*, 181–190.
9. Sistrunk, M.L.; Antosiewicz, D.M.; Purugganan, M.M.; Braam, J. Arabidopsis *TCH3* encodes a novel Ca^{2+} binding protein and shows environmentally induced and tissue-specific regulation. *Plant Cell* **1994**, *6*, 1553–1565. [CrossRef] [PubMed]
10. Sato, Y.; Kadota, A.; Wada, M. Mechanically induced avoidance response of chloroplasts in fern protonemal cells. *Plant Physiol.* **1999**, *121*, 37–44. [CrossRef] [PubMed]
11. Erner, Y.; Jaffe, M.J. Thigmomorphogenesis: The involvement of auxin and abscisic acid in growth retardation due to mechanical perturbation. *Plant Cell Physiol.* **1982**, *23*, 935–941.
12. Wang, L.; Weller, C.L. Recent advances in extraction of nutraceuticals from plants. *Trends Food Sci. Technol.* **2006**, *17*, 300–312. [CrossRef]
13. Hou, T.Z.; Li, M.D. Experimental evidence of a plant meridian system: IV. The effects of acupuncture on growth and metabolism of *Phaseolus vulgaris* L. beans. *Am. J. Chin. Med.* **1997**, *25*, 135–142. [CrossRef] [PubMed]

14. Hou, T.Z.; Luan, J.Y.; Wang, J.Y.; Li, M.D. Experimental evidence of a plant meridian system: III. The sound characteristics of Phylodendron (Alocasia) and effects of acupuncture on those properties. *Am. J. Chin. Med.* **1994**, *22*, 205–214. [CrossRef] [PubMed]

15. Hou, T.Z.; Re, Z.W.; Li, M.D. Experimental evidence of a plant meridian system: II. The effects of needle acupuncture on the; temperature changes of soybean (*Glycine max*). *Am. J. Chin. Med.* **1994**, *22*, 103–110. [CrossRef] [PubMed]

16. Yang, X.C.; Wang, B.C.; Duan, C.R.; Dai, C.Y.; Jia, Y.; Wang, X.J. Brief study on physiological effects of sound field on actinidia Chinese callus. *J. Chongqing Univ.* **2002**, *25*, 79–84. (In Chinese)

17. Yang, X.C.; Wang, B.C.; Duan, C.R. Effects of sound stimulation on energy metabolism of Actinidia chinensis callus. *Colloids Surf. B Biointerfaces* **2003**, *30*, 67–72.

18. Yang, X.C.; Wang, B.C.; Ye, M. Effects of different sound intensities on root development of Actinidia Chinese plantlet. *Chin. J. Appl. Environ. Biol.* **2004**, *10*, 274–276. (In Chinese)

19. Taş, F. Internal Acoustic Research and Improvement Methods in a Commercial Vehicle. Ph.D. Thesis, Istanbul Technical University, Istanbul, Turkey, 2010.

20. Hoagland, D.R.; Arnon, D.I. The water culture method for growing plants without soil. In *Circular California Agricultural Experiment Station*; University of California: Berkeley, CA, USA, 1938; pp. 347–461.

21. Kafkas, E.; Koşar, M.; Paydaş, S.; Kafkas, S.; Başer, K.H.C. Quality characteristics of strawberry genotypes at different maturation stages. *Food Chem.* **2007**, *100*, 1229–1236. [CrossRef]

22. Pinheiro, C.; Chaves, M.M. Photosynthesis and drought: Can we make metabolic connections from available data? *J. Exp. Bot.* **2010**, *62*, 869–882. [CrossRef] [PubMed]

23. Cramer, G.R.; Ergül, A.; Grimplet, J.; Tillett, R.L.; Tattersall, E.A.; Bohlman, M.C.; Quilici, D. Water and salinity stress in grapevines: Early and late changes in transcript and metabolite profiles. *Funct. Integr. Genom.* **2007**, *7*, 111–134. [CrossRef] [PubMed]

24. Kilian, J.; Whitehead, D.; Horak, J.; Wanke, D.; Weinl, S.; Batistic, O.; Harter, K. The AtGenExpress global stress expression data set: Protocols, evaluation and model data analysis of UV-B light, drought and cold stress responses. *Plant J.* **2007**, *50*, 347–363. [CrossRef] [PubMed]

25. Hasegawa, P.M.; Bressan, R.A.; Zhu, J.K.; Bohnert, H.J. Plant cellular and molecular responses to high salinity. *Annu. Rev. Plant Biol.* **2000**, *51*, 463–499. [CrossRef] [PubMed]

26. Munns, R. Comparative physiology of salt and water stress. *Plant Cell Environ.* **2002**, *25*, 239–250. [CrossRef] [PubMed]

27. Rosa, S.B.; Caverzan, A.; Teixeira, F.K.; Lazzarotto, F.; Silveira, J.A.; Ferreira-Silva, S.L.; Abreu-Neto, J.; Margis, R.; Margis-Pinheiro, M. Cytosolic APx knockdown indicates an ambiguous redox responses in rice. *Phytochemistry* **2010**, *71*, 548–558. [CrossRef] [PubMed]

28. Britt, A.B. Repair of DNA damage induced by solar UV. *Photosynth. Res.* **2004**, *81*, 105–112. [CrossRef]

29. Hilal, M.; Parrado, M.F.; Rosa, M.; Gallardo, M.; Orce, L.; Massa, E.M.; Prado, F.E. Epidermal lignin deposition in quinoa cotyledons in response to UV-B radiation. *Photochem. Photobiol.* **2004**, *79*, 205–210. [CrossRef]

30. Ibañez, S.; Rosa, M.; Hilal, M.; González, J.A.; Prado, F.E. Leaves of Citrus aurantifolia exhibit a different sensibility to solar UV-B radiation according to development stage in relation to photosynthetic pigments and UV-B absorbing compounds production. *J. Photochem. Photobiol. B Biol.* **2008**, *90*, 163–169. [CrossRef] [PubMed]

31. Bartwal, A.; Mall, R.; Lohani, P.; Guru, S.K.; Arora, S. Role of secondary metabolites and brassinosteroids in plant defense against environmental stresses. *J. Plant Growth Regul.* **2013**, *32*, 216–232. [CrossRef]

32. Seigler, D.S. *Plant Secondary Metabolism*; Kluwer Academic Publishers: Boston, MA, USA, 1998.

33. DiCosmo, F.; Misawa, M. Eliciting secondary metabolism in plant cell cultures. *Trends Biotechnol.* **1985**, *3*, 318. [CrossRef]

34. Sudha, G.; Ravishankar, G.A. Influence of methyl jasmonate and salicylic acid in the enhancement of capsaicin production in cell suspension cultures of *Capsicum frutescens* Mill. *Curr. Sci.* **2003**, *85*, 1212–1217.

35. Karuppusamy, S. A review on trends in production of secondary metabolites from higher plants by in vitro tissue, organ and cell cultures. *J. Med. Plants Res.* **2009**, *3*, 1222–1239.

36. Dixon, R.A.; Paiva, N.L. Stress-induced phenylpropanoid metabolism. *Plant Cell* **1995**, *7*, 1085. [CrossRef] [PubMed]

37. Cook, D.; Fowler, S.; Fiehn, O.; Thomashow, M.F. A prominent role for the CBF cold response pathway in configuring the low-temperature metabolome of Arabidopsis. *Proc. Natl. Acad. Sci. USA* **2004**, *101*, 15243–15248. [CrossRef] [PubMed]

38. Chalker-Scott, L.; Fnchigami, L.H. The role of phenolic compounds in plant stress responses. In *Low Temperature Stress Physiology in Crops*; Paul, H.L., Ed.; CRC Press Inc.: Boca Raton, FL, USA, 1989; p. 40.

39. Kaplan, F.; Kopka, J.; Sung, D.Y.; Zhao, W.; Popp, M.; Porat, R.; Guy, C.L. Transcript and metabolite profiling during cold acclimation of Arabidopsis reveals an intricate relationship of cold-regulated gene expression with modifications in metabolite content. *Plant J.* **2007**, *50*, 967–981. [CrossRef] [PubMed]

40. Wingler, A.; Roitsch, T. Metabolic regulation of leaf senescence: Interactions of sugar signalling with biotic and abiotic stress responses. *Plant Biol.* **2008**, *10*, 50–62. [CrossRef] [PubMed]

41. Hare, P.D.; Cress, W.A.; Van Staden, J. Dissecting the roles of osmolyte accumulation during stress. *Plant Cell Environ.* **1998**, *21*, 535–553. [CrossRef]

42. Avonce, N.; Leyman, B.; Mascorro-Gallardo, J.O.; Van Dijck, P.; Thevelein, J.M.; Iturriaga, G. The Arabidopsis trehalose-6-P synthase AtTPS1 gene is a regulator of glucose, abscisic acid, and stress signalling. *Plant Physiol.* **2004**, *136*, 3649–3659. [CrossRef] [PubMed]

43. Pego, J.V.; Kortstee, A.J.; Huijser, C.; Smeekens, S.C.M. Photosynthesis, sugars and the regulation of gene expression. *J. Exp. Bot.* **2000**, *51*, 407–416. [CrossRef] [PubMed]

44. Ramon, M.; Rolland, F.; Sheen, J. Sugar sensing and signaling. The Arabidopsis Book. *Am. Soc. Plant Biol.* **2008**. [CrossRef]

45. Van den Ende, W.; Valluru, R. Sucrose, sucrosyl oligosaccharides, and oxidative stress: Scavenging and salvaging? *J. Exp. Bot.* **2009**, *60*, 9–18. [CrossRef] [PubMed]

46. Bolouri-Moghaddam, M.R.; Le Roy, K.; Xiang, L.; Rolland, F.; Van den Ende, W. Sugar signalling and antioxidant network connections in plant cells. *FEBS J.* **2010**, *277*, 2022–2037. [CrossRef] [PubMed]

47. Kalt, W.; Ryan, D.A.J.; Duy, J.C.; Prior, R.L.; Ehlenfeldt, M.K.; Vander Kloet, S.P. Interspecific variation in anthocyanins, phenolics, and antioxidant capacity among genotypes of highbush and lowbush blueberries (*Vaccinium section cyanococcus* spp.). *J. Agric. Food Chem.* **2001**, *49*, 4761–4767. [CrossRef] [PubMed]

48. Lee, S.K.; Kader, A.A. Preharvest and postharvest factors influencing vitamin C content of horticultural crops. *Postharvest Biol. Technol.* **2000**, *20*, 207–220. [CrossRef]

49. Dumas, Y.; Dadomo, M.; Di Lucca, G.; Grolier, P. Effects of environmental factors and agricultural techniques on antioxidantcontent of tomatoes. *J. Sci. Food Agric.* **2003**, *83*, 369–382. [CrossRef]

50. Bartoli, C.G.; Yu, J.; Gomez, F.; Fernández, L.; McIntosh, L.; Foyer, C.H. Inter-relationships between light and respiration in the control of ascorbic acid synthesis and accumulation in Arabidopsis thaliana leaves. *J. Exp. Bot.* **2006**, *57*, 1621–1631. [CrossRef] [PubMed]

51. Smirnoff, N.; Conklin, P.L.; Loewus, F.A. Biosynthesis of ascorbic acid in plants: A renaissance. *Annu. Rev. Plant Biol.* **2001**, *52*, 437–467. [CrossRef] [PubMed]

52. Pallanca, J.E.; Smirnoff, N. The control of ascorbic acid synthesis and turnover in pea seedlings. *J. Exp. Bot.* **2000**, *51*, 669–674. [CrossRef] [PubMed]

53. Green, M.A.; Fry, S.C. Vitamin C degradation in plant cells via enzymatic hydrolysis of 4-*O*-oxalyl-L-threonate. *Nature* **2005**, *433*, 83–87. [CrossRef] [PubMed]

54. Horemans, N.; Foyer, C.H.; Asard, H. Transport and action of ascorbate at the plant plasma membrane. *Trends Plant Sci.* **2000**, *5*, 263–267. [CrossRef]

55. Ferrandino, A.; Lovisolo, C. Abiotic stress effects on grapevine (*Vitis vinifera* L.): Focus on abscisic acid-mediated consequences on secondary metabolism and berry quality. *Environ. Exp. Bot.* **2014**, *103*, 138–147. [CrossRef]

56. Gill, S.S.; Tuteja, N. Reactive oxygen species and antioxidant machinery in abiotic stress tolerance in crop plants. *Plant Physiol. Biochem.* **2010**, *48*, 909–930. [CrossRef] [PubMed]

57. Poiroux-Gonord, F.; Bidel, L.P.; Fanciullino, A.L.; Gautier, H.; Lauri-Lopez, F.; Urban, L. Health benefits of vitamins and secondary metabolites of fruits and vegetables and prospects to increase their concentrations by agronomic approaches. *J. Agric. Food Chem.* **2010**, *58*, 12065–12082. [CrossRef] [PubMed]

58. Grassmann, J.; Hippeli, S.; Elstner, E.F. Plant's defence and its benefits for animals and medicine: Role of phenolics and terpenoids in avoiding oxygen stress. *Plant Physiol. Biochem.* **2002**, *40*, 471–478. [CrossRef]

59. Kunz, D.A.; Chen, J.L.; Pan, G. Accumulation of α-keto acids as essential components in cyanide assimilation by Pseudomonas fluorescens NCIMB 11764. *Appl. Environ. Microbiol.* **1998**, *64*, 4452–4459. [PubMed]

60. English-Loeb, G.; Stout, M.J.; Duffey, S.S. Drought stress in tomatoes: Changes in plant chemistry and potential nonlinear consequences for insect herbivores. *Oikos* **1997**, *79*, 456–468. [CrossRef]

61. Fujita, M.; Fujita, Y.; Noutoshi, Y.; Takahashi, F.; Narusaka, Y.; Yamaguchi-Shinozaki, K.; Shinozaki, K. Crosstalk between abiotic and biotic stress responses: A current view from the points of convergence in the stress signaling networks. *Curr. Opin. Plant Biol.* **2006**, *9*, 436–442. [CrossRef] [PubMed]

agronomy

MDPI

Review

Molecular and Genetic Bases of Fruit Firmness Variation in Blueberry—A Review

Francesco Cappai †, Juliana Benevenuto †, Luís Felipe V. Ferrão † and Patricio Munoz *

Blueberry Breeding and Genomics Laboratory, Horticultural Sciences Department, University of Florida, Gainesville, FL 32611, USA; francesco.cappai@ufl.edu (F.C.); jbenevenuto@ufl.edu (J.B.); lferrao@ufl.edu (L.F.V.F.)
* Correspondence: p.munoz@ufl.edu; Tel.: +1-352-273-4837
† These authors contributed equally to this work.

Received: 3 August 2018; Accepted: 30 August 2018; Published: 5 September 2018

Abstract: Blueberry (*Vaccinium* spp.) has been recognized worldwide as a valuable source of health-promoting compounds, becoming a crop with some of the fastest rising consumer demand trends. Fruit firmness is a key target for blueberry breeding as it directly affects fruit quality, consumer preference, transportability, shelf life, and the ability of cultivars to be machine harvested. Fruit softening naturally occurs during berry development, maturation, and postharvest ripening. However, some genotypes are better at retaining firmness than others, and some are crispy, which is a putatively extra-firmness phenotype that provides a distinct eating experience. In this review, we summarized important studies addressing the firmness trait in blueberry, focusing on physiological and molecular changes affecting this trait at the onset of ripening and also the genetic basis of firmness variation across individuals. New insights into these topics were also achieved by using previously available data and historical records from the blueberry breeding program at the University of Florida. The complex quantitative nature of firmness in an autopolyploid species such as blueberry imposes additional challenges for the implementation of molecular techniques in breeding. However, we highlighted some recent genomics-based studies and the potential of a QTL (Quantitative Trait Locus) mapping analysis and genome editing protocols such as CRISPR/Cas9 to further assist and accelerate the breeding process for this important trait.

Keywords: firmness; *Vaccinium*; ripening; cell wall; crispy; quantitative genetics; breeding; molecular markers; genome editing

---

1. Introduction

Blueberry has been recognized worldwide for its health benefits due to its high content and wide diversity of polyphenolic compounds [1,2]. Polyphenolic compounds, especially anthocyanins, have been shown to have anti-oxidant, anti-inflammatory, anti-proliferative, anti-obesity, and neuroprotective properties [3]. Such health-related awareness has been driving an increase in demand for blueberries, which became one of the crops with the highest production trends. From 1996 to 2016, worldwide production has grown by 72.11% [4]. Currently, the United States of America is the largest producer, being responsible for around 48% of the world's production in 2016 [4].

Most of the blueberry production is destined for the fresh market sector, which requires high-quality berries and postharvest longevity [5]. In order to maintain the fresh market standards, blueberries are usually hand-picked. However, hand-harvest labor accounts for 50% (or even more) of the production costs and raises concerns about its long-term availability [5–7]. In addition, fresh fruit handling by ill workers has been linked to foodborne illnesses in consumers [8]. Mechanical harvesting can mitigate the need of hand-harvest labor, decrease production costs, and foster further expansion of this healthy fruit. Wide adoption of machine harvesting for the fresh market is, however, currently limited to only a few commercial blueberry cultivars. Fruit firmness is one of the main determinant traits required to withstand the physical impacts during machine harvesting [6,7,9–11].

Blueberries are shipped to long-distance markets all over the world and softening and bruising fruits are among the most common defects causing shipment rejections [10,12]. Hence, firm berries are also critical for transportability and shelf-life longevity [10,11]. Moreover, a firm texture is one of the consumers' most appreciated features, being associated with the general concept of fruit freshness and quality [13–15]. Taken all together, fruit firmness is a key target for blueberry breeding as it benefits all stakeholders.

Firmness naturally declines during berry development, maturation, and postharvest ripening [12,16,17]. Understanding the biological mechanisms underlying fruit softening is essential to manipulate it without affecting other desirable aspects of ripening, such as color, flavor, aroma, or nutritional value [18]. Despite the natural softening during ripening, some plant genotypes appeared best at retaining firmness by the merit of having high initial firmness values [9]. Intraspecific (*Vaccinium corymbosum* L.) and interspecific (wild relatives) phenotypic variation has been reported, constituting breeding resources for firmness improvement [6,19]. In addition to firmness, a potentially separate phenomenon has been identified in blueberry: fruit crispiness. Crispy fruits are firmer and have higher bursting energy than standard fruits. Hence, crispiness is a distinct textural attribute that is also of high interest for the blueberry industry.

In this review, we presented recent advances in blueberry research regarding firmness changes during berry development and ripening. We summarized the extent of phenotypic variability of the firmness trait described in blueberry and its genetic parameters. New insights into these topics were achieved by using previously available data and historical records from the blueberry breeding program at the University of Florida. We also emphasized the first study attempting to implement marker-assisted selection (MAS) for this trait, through genome-wide association studies (GWAS). Finally, we discussed perspectives to further implement QTL mapping and genome editing technologies, such as CRISPR in blueberry.

2. Physiological, Cellular, and Molecular Changes Affecting Fruit Firmness

2.1. Fruit Anatomy and Growth during Ripening

Blueberries, as many berries of the *Vaccinium* genus, are false berries, because they are berry-like, but develop into a fruit without a stone from a single fertilized superior ovary [20]. Anatomically, the blueberry fruit develops from an inferior ovary. The endocarp is composed of five carpels with ten locules and five lignified placentae to which around 10–65 seeds are attached. The endocarp is surrounded by the mesocarp, which is composed of parenchyma cells along with rings of vascular bundles, with some stone cells unevenly distributed [15,21]. The epicarp originates from the flower calyx. Delimited by a ring of vascular bundles, the hypodermal layer contains the anthocyanin pigmentation. Above this, there is a single layer of epidermis without stomates. The epidermis is covered by a cuticle and an epicuticular waxy bloom that overshadows the purple-black skin of blueberry fruits, creating the characteristic light-blue color [22].

Berry growth commonly exhibits a double sigmoid growth curve, with a lag stage between two phases of active growth [23,24]. The first growth stage occurs after syngamy and is characterized by a rapid cell division, leading to a rapid expansion of the pericarp. In the second stage (lag stage), embryo and endosperm tissues mature, while the development of the pericarp is retarded with no evident increase in berry size and no changes in color. In the third stage, a second rapid pericarp development takes place due to cell enlargement until the berry is fully ripe. Changes in size are accompanied by changes in color (from green, pink, to blue), and biochemical changes such as increases in pH, sugar composition, soluble solid content, volatiles, and texture. The fruit development and ripening process usually takes 45–90 days, depending on the cultivar and external factors [25].

2.2. Ripening-Associated Physiological Changes

Physiological changes associated with ripening are a major determinant of berry firmness, as the fruit softens as it matures. The ripening process is coordinated by a complex network of endogenous hormones. For many fleshy fruit plant species, the phytohormone ethylene is the main ripening agent [26]. Climacteric fruits show a concomitant increase in respiration and ethylene biosynthesis upon initiation of ripening [27]. Ethylene perception and signal transduction can affect fruit color, sugar and acid content, and firmness [28]. This process is however not linear: it varies over time and the abovementioned ripening changes can be disjointed [28,29]. Currently, there is no consensus on whether blueberry is a climacteric fruit or not. Some studies suggested that blueberry is indeed a climacteric fruit [30,31], with also increased anthocyanin accumulation upon ethylene application in some cultivars [32]. Chiabrando and Giacalone (2011) [33] proposed that 1-methylcyclopropene (1-MCP), a compound that hinders cellular ethylene perception, has the potential for controlling ripening in blueberry. Fruits from the cultivar "LateBlue" showed reduced post-harvest weight loss and lower total soluble solid content after 1-MCP treatment, indicating a slower ripening effect [33]. In line with these findings, Wang et al. (2018) [34] also showed that an ethylene absorbent treatment prevented ethylene production, inhibited cell wall degrading enzymes, and reduced softening of blueberry fruit. However, conversely, other studies suggested that 1-MCP application has no inhibition effect on ripening of post-harvested blueberries [35,36], and might decrease fruit firmness [37,38].

In non-climacteric fruits, where no burst in ethylene production is observed, abscisic acid (ABA) seems to have a stronger role during fruit ripening [27]. However, similarly to what has been reported for ethylene, studies addressing the effects of ABA in blueberry fruit firmness also showed contradictory results. Sun et al. (2013) [39] reported that ABA application was able to promote fruit softening; whereas Buran et al. (2012) [40] observed the opposite phenomenon.

As of now, the role of ethylene and ABA phytohormones in blueberry fruit ripening, as well as the question if blueberry is a climacteric or non-climacteric fruit remains unclear. However, it is noteworthy how multiple studies have reported different effects of ethylene, as inhibitor or promoter of fruit ripening, and also cultivar-specific responses, suggesting the need for more studies in this field and that distinct mechanisms might exist among blueberry genotypes.

2.3. Molecular and Architectural Changes in Plant Cell during Fruit Ripening

Ripening is also accompanied by compositional and architectural changes in the plant cell, mainly at the primary cell wall and middle lamella through the action of carbohydrate active enzymes, ultimately affecting fruit firmness [41]. Plant cell wall is a complex matrix of polysaccharides, mainly composed of pectin, hemicellulose, and cellulose. The cellulose microfibrils are cross-linked through hydrogen bonds with hemicellulose, thereby forming a complex network that provides tensile strength to the primary cell wall [42]. The pectin matrix is interwoven with the cellulose-hemicellulose network in the primary cell wall and also composes the middle lamella, being a major physical mediator of cell adhesion and separation [43].

Noncellulosic sugars have been implied in regulating firmness as they provide physical support by connecting at cellulose myofibrils in cell walls [44]. In particular, depolymerization, and solubilization of hemicellulose and pectin are the processes generally associated with cell wall disassembly and fruit softening during ripening [41,45–48]. In blueberry, hemicellulosic polymers undergo a significant depolymerization and a moderate solubilization throughout all five stages of ripeness analyzed, indicating that hemicellulose modification might be the main cell wall alteration during blueberry ripening [49]. The pectin matrix also shows increased solubilization mostly at the initial and intermediate stages of ripening and little reduction in polymer size occurred [49]. Pectin solubilization was also observed during blueberry post-harvest storage [17]. Interestingly, calcium application has been reported to enhance firmness and post-harvest quality of blueberries [50,51]. Angeletti et al. (2010) [51] speculated that the increase in calcium content might decrease pectin solubilization, while not affecting hemicellulose in blueberry. In many ripe fruits, most of the cell-to-cell adhesion is conferred by calcium-pectate cross-links in the middle lamella, producing a semi-rigid gel that increases cell wall stiffness and hinders cell wall disassembly [41,43,52,53]. In general, the role of calcium in cell wall solidity and post-harvest quality is well established in a number of crops, such as apple (*Malus domestica* Borkh) [54–56], grape (*Vitis vinifera* L.) [57–59], and strawberry (*Fragaria* × *ananassa* Duchesne) [60–62], and seems to be relevant also in blueberry.

The degradation of cell wall polysaccharides depends on the action of enzymes with distinct spectra of activities. A panel of enzymes involved in carbohydrate synthesis, modification, and breakdown are collectively named as carbohydrate-active enzymes (CAZymes). The CAZy database is currently the most comprehensive database (http://www.cazy.org) for CAZyme proteins [63]. Hence, to further investigate the role of CAZymes during blueberry ripening, we retrieved the putative encoding genes by sequence similarity of the predicted blueberry proteome with the CAZy database and the transcriptional profile of five stages of berry fruit development (pad, cup, green, pink, and ripe fruits) from Gupta et al. (2015) [64]. Distinct patterns of expression were observed for the genes into the six classes of CAZymes throughout the fruit development process (Figure 1). Up-regulated CAZymes encoding genes in pink and ripe fruits in relation to early stages are highlighted as they can be involved in cell wall disassembly, allowing cell expansion and fruit growth, but later cause fruit softening.

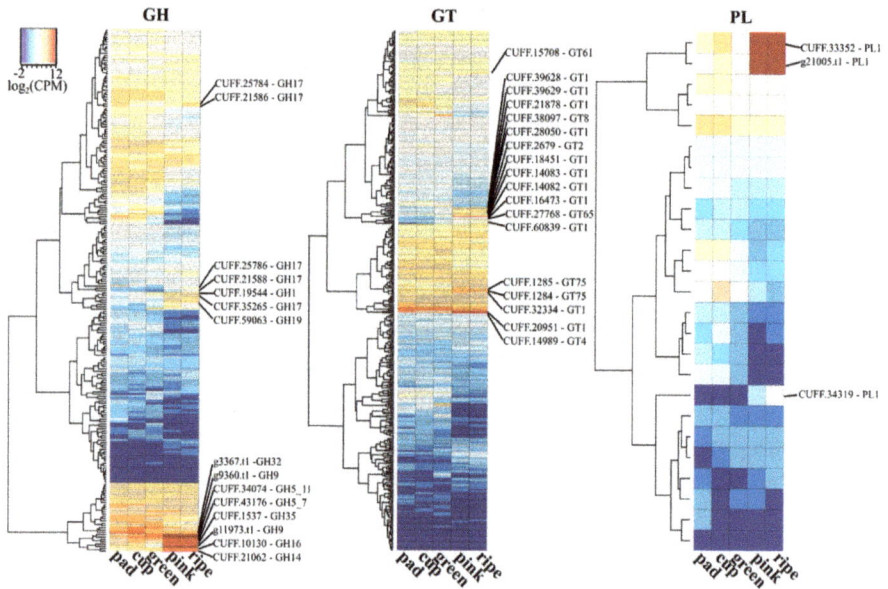

Figure 1. *Cont.*

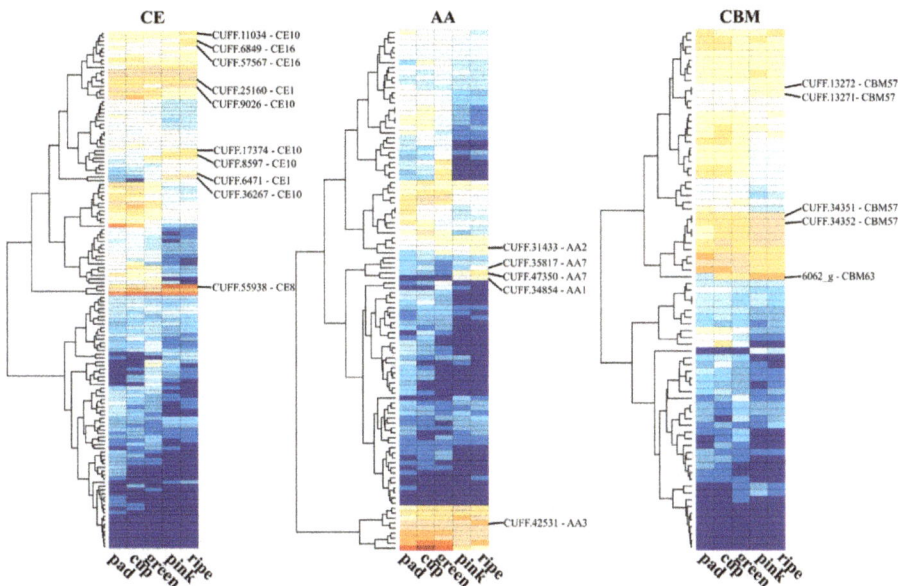

Figure 1. RNA-seq transcriptional profile of CAZymes during five stages of berry fruit development and ripening. The predicted blueberry proteome was screened for CAZymes [65] using Hmmscan from the HMMER v3.1b2 package (http://hmmer.org/) and the dbCAN HMM profile database [66]. The hmmscan-parser script provided by dbCAN was used to select significant matches. CPM (counts per million mapped reads) values from five developmental berry stages of a RNA-seq experiment were retrieved from Gupta et al. (2015) [64] from the bitbucket repository (https://bitbucket.org/lorainelab/blueberrygenome). The CPM mean value of each gene was calculated from three replicates of each stage available. The $\log_2$-transformed CPM mean values were used to plot the gene expression of CAZymes across the five stages. The R package "gplots" with "heatmap.2" function was used to generate the heatmaps using the parameter "dendrogram = c("row")" to cluster genes according to hierarchical clusters. Classes and modules of CAZymes are represented by GH (glycoside hydrolases), GT (glycosyl transferases), PL (polysaccharide lyases), CE (carbohydrate esterases), AA (auxiliary activities), and CBM (carbohydrate-binding modules) (For more details, see: http://www.cazy.org/). Genes up-regulated at pink and/or ripe stages in relation to early stages are highlighted and the family number reported.

Glycoside hydrolases (GHs) catalyze the hydrolysis of glycosidic bonds between carbohydrates or between a carbohydrate and a non-carbohydrate moiety. Among the up-regulated GHs families, we found five beta-glucosidases (GH family 17) blueberry encoding genes (Figure 1). GH 17 family members have been associated with cell wall degradation and remodeling, hydrolysis of phytohormones (such as ABA), pathogen resistance, and in aromatic acid biosynthesis pathways [67]. In banana (*Musa* spp.), more than a two-fold enhanced expression of genes in the GH 17 family was detected in the ripe fruits compared to unripe ones [67]. Others interesting GH enzymes detected were endo-1,4-β-glucanases (GH7, GH9), exo-1,4-β-glucanases (GH1), endo-xyloglucan transferase (GH16), and endo-1,4-β-xylanase (GH11) that probably act on hemicellulose depolymerization or hydrolyzing the crosslinks between microfibrils, thereby disassembling the cellulose–xyloglucan network [68,69]. Exo-β-galactosidases from GH family 35 has also been associated with depolymerization of pectins. These enzymes can cause cell wall loosening during growth, and cell wall degradation during ripening and senescence [67,70]. Up-regulated invertase (GH32) and amylase (GH14) generate reducing sugar monomers, which may be responsible for the increased sweetness during fruit ripening [71].

In blueberry, cold storage was shown to maintain firmness probably due to the reduced activities of such enzymes at lower temperatures [36].

Polysaccharide lyases (PLs) cleave uronic acid-containing polysaccharide chains. Three pectate lyase (PL1 family) genes were up-regulated during blueberry ripening (Figure 1). PL1 enzymes degrades de-esterified pectin in the primary wall. In tomato (*Solanum lycopersicum* L.), the silencing of a PL-encoding gene inhibited pectin solubilization and depolymerization, maintaining the firmness without affecting other aspects of ripening [18,72]. In line with this, a gene in the carbohydrate esterase (CEs) class, encoding a pectin methylesterase (CE8), was also up-regulated in blueberry (Figure 1). CE8 enzymes alter the pectin structure by catalyzing the demethyl esterification of pectin, which then becomes cleavable by pectate lyases [18]. Glycosyltransferases (GTs), which catalyze the transfer of sugar moieties forming glycosidic bonds, have been associated with cell wall polysaccharide synthesis (GT2, GT8, GT61, GT65, GT75), sucrose synthesis (GT4), and anthocyanin modification and formation of glycosylated volatile compounds during fruit development and ripening [73,74]. Enzymes with auxiliary activities (AAs) that act in conjunction with CAZymes and genes with a carbohydrate-binding module were also found up-regulated at pink and blue stages of blueberry ripening (Figure 1).

In addition to changes in cell wall composition and disassembly, reduction in turgor pressure during ripening can also cause shrinkage and loss of firmness. Reduction in the turgor is likely due to the accumulation of osmotic solutes in the apoplast and to water loss by the fruit [41] and has been reported in stored blueberries [75]. Recent studies have also shown the roles of the stem scar size/transpiration and cuticular wax composition and thickness in water loss and maintenance of the post-harvest firmness in blueberry [22,76,77]. Among the blueberry fruit cuticular triterpenoid composition, the ursolic acid content at harvest was positively correlated to weight loss and softening after storage, offering an interesting target for further studies [77]. The importance of containing water loss is also supported by the fact that coating fruits with oily films, which create a semi-permeable barrier around the fruit, prevented firmness decay [78,79].

The findings reported herein offer some insights into the molecular bases of fruit softening during ripening and postharvest storage of blueberries. Further functional validation of the differentially expressed CAZymes can provide good candidates for the implementation of genome editing tools aiming at firmness maintenance.

2.4. Fruit Tissue and Cellular Differences Underlying Firmness Variation

Several components of fruit tissues have been reported to contribute to firmness and texture variation across genotypes, including cell type, size, shape, packing, cell-to-cell adhesion, extracellular space, and cell wall thickness [15,80,81]. In blueberry, differences regarding the number and organization of lignified cells with thick secondary cell walls, such as stone cells in the mesocarp, have been hypothesized to strengthen the surrounding flesh tissue and contribute to fruit firmness variation among genotypes [81,82]. Sensorial studies of genotypes ranging from soft to crisp suggested that the crisp texture may be related to the blueberry skin toughness rather than the flesh [83]. Histological analyses of cell type, area, and structure of the outermost cell layers of genotypes varying in textural attributes showed that crispy genotypes had a smaller average cell area compared with standard-textured genotypes in mature fruits and no difference in the frequency of stone cells in the layers beneath the epidermis was detected [15].

3. Genetics and Breeding of Blueberry Firmness

3.1. Measuring Firmness

Firmness has also been a key breeding target for many fruit crops such as apple, pear (*Pyrus communis* L.), peaches (*Prunus persica* L. Batsch), etc. Historically, firmness in these crops has been measured using a hand-held penetrometer [84–88] (e.g., Magness-Taylor's and Effe-gi testers) and, more recently, an Instron-mounted probe (Instron Corporation, Norwood, MA USA) [89–92].

A number of attempts has been undertaken to upscale firmness measurement in these crops through imaging and acoustic studies, with partially successful results [93–98]. However, these methods were not translated to blueberry, possibly due to the constraints imposed by the small size of the blueberry fruit compared to pear or apple.

In blueberry, fruit firmness has been traditionally evaluated by chewing texture, with the assignment of a subjective and qualitative score ranging from "soft" to "firm" [99]. More recently, automated equipment allowed the objective and quantitative evaluation of fruit firmness. FirmTech instruments (FirmTech I and II, Bioworks, Wamego, KS, USA) are the most commonly used for blueberry and firmness is a measure of the compression force (g) required to deflect the surface of the fruit one millimeter. A texture analyzer instrument (Texture Technologies Corporation, Scardale, NY, USA) has also been increasingly used and might give more precise results than FirmTech, especially when trying to dissect skin and pulp firmness. However, these equipment present differences in probe sizes and shapes, making the standardization challenging [83,100–103]. Hence, in this review, we only included measurements obtained with FirmTech instruments as more data were available [6,14,37,80,88–138].

Regarding the firmness trait evaluation, an issue that has not been addressed yet is how many berries should be sampled to accurately represent a genotype. Each study has used a distinct number of samples, ranging from five to hundreds of berries. Assuming that sample size is an important element for research design and validation, we assessed how the mean firmness varied across different sample sizes considering four cultivars (Figure 2). We collected 200 berries from each genotype and performed 1000 resamplings considering 12 different sample sizes. The mean and variance values computed for the 200 berries were considered our "true" population parameters. As expected, small sample sizes resulted in higher variance in the mean firmness values of a genotype than large sample sizes (Figure 2a). The fraction of the confidence intervals (CI = 90%) that encompassed the "true" population mean ranged across the sample sizes for all cultivars (Figure 2b) and, the smaller the sample size, the more values out of the CI were generated. For a sample size of five berries and for all cultivars, we observed that around 25% of the estimates did not lie in the confidence interval of 90%, which means that in a breeding population composed of 1000 individuals around of 250 measures of firmness would be inaccurate, hence affecting downstream analyses. By sampling 25 berries, the amount of outlier estimates dropped to about 10% considering the confidence interval and it is also a sample size operationally feasible, since FirmTech devices are able to run 25 berries per round of analysis.

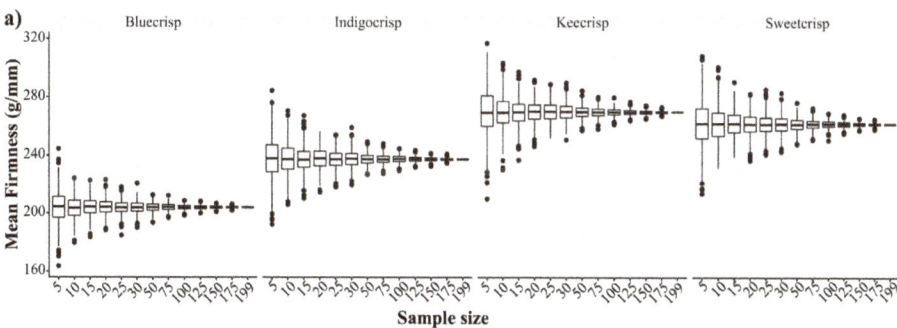

Figure 2. *Cont.*

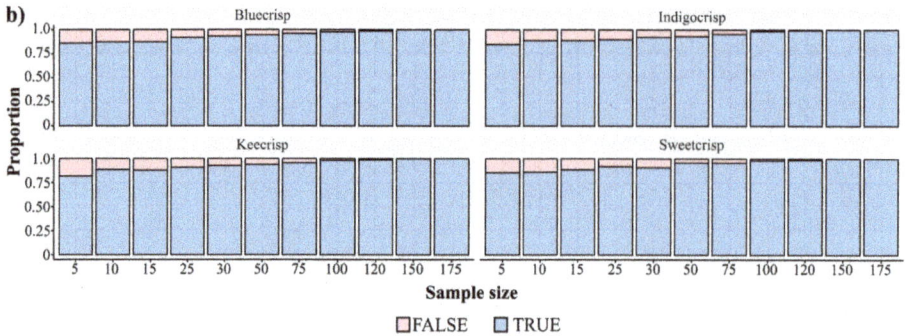

Figure 2. Number of berry samples to estimate the firmness value of a genotype. (**a**) Boxplot of the mean firmness values considering different sample sizes for four SHB cultivars. Briefly, 200 mature berries of each cultivar were collected and measured for firmness using the FirmTech II device (BioWorks Inc., Victor, NY, USA). To describe the effect of the sample size on the mean value, 1000 resamplings were carried out considering 12 different sample sizes (5–199) and the mean firmness value was computed at each sampling round. (**b**) Fraction of the confidence intervals that encompassed the "true" mean parameter, computed assuming that 200 berries are enough to well-represent a genotype, across different sample sizes and four different cultivars. We performed the same resampling process but recording the parameters that laid in a confidence level of 90%.

3.2. Phenotypic Variation and Breeding Improvement in Fruit Firmness

Heritable phenotypic variation is critical for breeding selection to be effective. Enough phenotypic variation for firmness has been observed within breeding populations, among blueberry cultivars, and wild species [6,19,139]. Herein, we performed a survey of firmness values for a wide range of cultivars, representing the main cultivated blueberry types. Cultivated blueberry belongs to section *Cyanococcus* of the genus *Vaccinium* into the Ericaceae Family and comprises distinct species and hybrids with specific ploidy, plant architecture, and chilling hour requirements. The main blueberry type used for commercial production is a tetraploid highbush (2n = 4X = 48), with *Vaccinium corymbosum* L. as the primary species in its genetic background. Highbush blueberries are further classified according to their chilling requirements as Northern Highbush Blueberry (NHB) and Southern Highbush Blueberry (SHB) [140,141]. NHB is native to Eastern North America and is the most widely planted type in temperate climates (600–1200 h of chilling between 0 to 7 °C). SHB was originally developed by the introgression of an evergreen Florida native species (*Vaccinium darrowii* Camp) into the NHB background, leading to reduced chilling requirements (100 to 600 h) and, therefore, being adapted to warmer climates such as southern US [142–144]. Rabbiteye blueberry is another commercially important species, hexaploid *Vaccinium ashei* Reade, which tolerates a range of soil and warm climatic conditions and are also planted in the Southern US. Lowbush blueberry is another tetraploid species (*Vaccinium angustifolium* Aiton), whose wild fruits are harvested commercially in New England. The cultivated half-high blueberry (HH) was originated from a cross between NHB and lowbush and also requires lower temperatures to flower.

In this survey, we collected berry firmness data from scientific papers that used the FirmTech instrument, except for those reporting experimental treatments such as insect damage, extreme growth conditions, etc. (see Supplementary Materials for more details). The mean firmness value of each cultivar was plotted according to their respective release year (Figure 3). High diversity of firmness values can be observed between and within blueberry types (Figure 3a). As the NHB breeding began earlier (1906), more NHB cultivars are displayed. The first released NHB cultivars, "Harding" and "Rubel", were actually selected from the wild. "Pioneer" was the first cultivar released after breeding efforts in 1920 [145]. SHB breeding programs are more recent and the first cultivar "Sharpblue" was released in 1975 [146]. The NHB cultivar "Herbert" presented the lowest firmness value, while the SHB cultivar "Sweetcrisp" presented the highest firmness value among all blueberry types (Table 1).

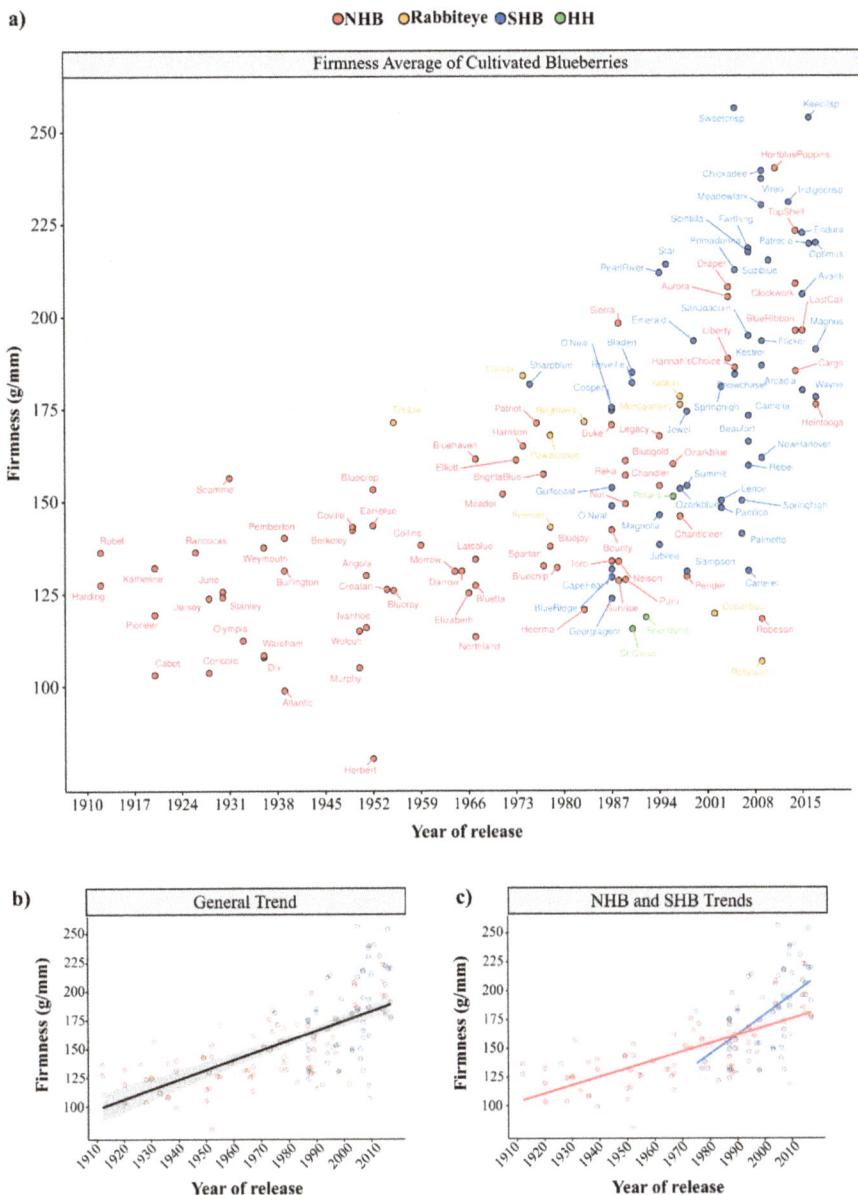

Figure 3. (**a**) Blueberry cultivar firmness values by year of release. Plot was created using historical data recorded by the blueberry breeding program at the University of Florida and data from previous blueberry publications (for details, see Supplementary Material); (**b**) Linear regression on the firmness values as a function of the year of cultivar release with a slope of 0.84; (**c**) Linear regression on the firmness values as a function of the year of cultivar release for NHB (slope of 0.72) and SHB (slope of 1.71), separately.

Table 1. Summary of firmness measurements for distinct types of cultivated blueberry. For more detailed information, see Supplementary Materials.

Type *	n	Mean	St. Dev	Maximum	Minimum
SHB	50	183	35.18	256 (Sweetcrisp)	124 (Georgiagem)
HH	3	128	19.79	151 (Polaris)	115 (St. Cloud)
NHB	74	145	31.22	240 (Hortblue Poppins)	80 (Hebert)
Rabbiteye	9	157	27.87	184 (Climax)	106 (Robeson)

* Southern Highbush (SHB), Northern Highbush (NHB), Half-high (HH), and Rabbiteye cultivars.

The positive regression slope in Figure 3b showed a general trend towards firmness improvement throughout the years, i.e., on average, modern cultivars are firmer than first-released genotypes. This general trend is expected, as firmness is a key target trait in blueberry breeding programs and also suggested that significant genetic gain has been achieved. In Figure 3c, we can also observe a trend towards higher firmness values of SHB in relation to NHB, although SHB also exhibit slighter higher variation (Table 1). Some studies suggested that cultivars with a higher percentage of *V. darrowii* (evergreen blueberry) and *V. ashei* (rabbiteye) ancestry often possessed higher firmness values, which would be the case of SHB; while cultivars with *V. angustifolium* (lowbush) ancestry presented softer fruits than the average, as also observed for HH blueberries in Figure 3a [6,99,139]. However, such conclusions can be misleading since the firmness values retrieved herein were collected under distinct experimental conditions and locations, using varying numbers of berries and equipment versions (see Supplementary Materials).

It is also noteworthy that cultivars considered crispy, such as "Sweetcrisp" (SHB), "Keecrisp" (SHB), "Indigocrisp" (SHB), "Hortblue Poppins" (NHB), showed the highest firmness values (Figure 3a). Blaker et al. (2014) [107] had also found a correlation among compression and bioyield force measures with sensory scores for bursting energy, flesh firmness, and skin toughness, distinguishing crisp from standard-texture SHB genotypes. We should also mention that in food science jargon crunchiness and crispiness are two different phenomena based on sensory, acoustic and vibrational cues [147]. However, no study has formally addressed whether blueberry cultivars are crispy or crunchy.

3.3. Quantitative Genetics of Firmness Trait in SHB Blueberry Breeding

Blueberry breeding programs have relied on pedigree information and cross-pollination breeding methods to increase the mean phenotypic performance in selected populations [141]. Phenotypic recurrent selection has been used as a primary breeding strategy in blueberry, where elite parents are selected at each generation for intercrossing, the progenies are evaluated over the course of multiple years in the field, and a cultivar consists of a superior individual cloned by cutting propagation. During this process, quantitative genetics constitute the fundamental basis to guide breeding efforts. Many agronomically important traits, including firmness, display a continuous phenotypic variation in a given population, likely governed by the joint action of numerous genes and environmental factors. The continuous distribution of such complex traits resulted in phenotypes that do not show simple Mendelian inheritance, requiring a quantitative genetics framework of analyses. Quantitative models dissect the observed phenotypic value of an individual into genetic and environmental components in order to estimate the heritable and non-heritable portions of the variation.

Quantitative genetic studies in blueberry have focused on three main challenges: (I) identify the elite genotypes to be used as parents in future crosses [19,148–151]; (II) estimate genetic parameters such as heritability, phenotypic and genetic correlations, and predict the expected change in a trait in response to selection—the breeder's equation [19,148] and (III) estimate the genotype-by-environment (G×E) interaction [13,19,152,153]. Genetics studies in blueberry have been reported for distinct traits; however, to our knowledge, only Cellon et al. (2018) [19] have investigated the firmness trait. In order to guide the selection of firmer genotypes and estimate genetic parameters, Cellon et al. (2018) [19]

found moderate-to-high values of heritability for this trait with large variability across the years (0.43 in 2014 and 0.70 in 2015). The presence of variability across the years indicated genotype-by-year interaction, while the magnitude of the heritability values suggested potentially rapid gains in response to selection.

Phenotypic expression of quantitative traits is influenced by environmental conditions, leading to variations across locations and years that, ultimately, impact the ranking of the genotypes. Despite its relevance, there are few studies addressing G×E for the firmness trait in the blueberry literature. Motivated by this, we performed an initial assessment of G×E interaction in SHB using the dataset recorded by the breeding program at the University of Florida for six cultivars, planted in four locations, from 2011 to 2017 (Figure 4a). Based on an analysis of variance (ANOVA), the G×E interaction effect was statistically significant ($p < 0.05$). The interaction plot suggests that the crispy cultivars, "Indigocrisp" and "Sweetcrisp", are more stable across the environments than the non-crispy ones (Figure 4b). This preliminary screening suggests that G×E interaction impacts fruit firmness. From a theoretical point of view, ignoring the heterogeneity caused by different environments can bias the prediction of breeding values and negatively affect the estimates of genetic variances. For practical purposes, such effect directly affects the selection of elite materials and decreases genetic gains along breeding cycles [154].

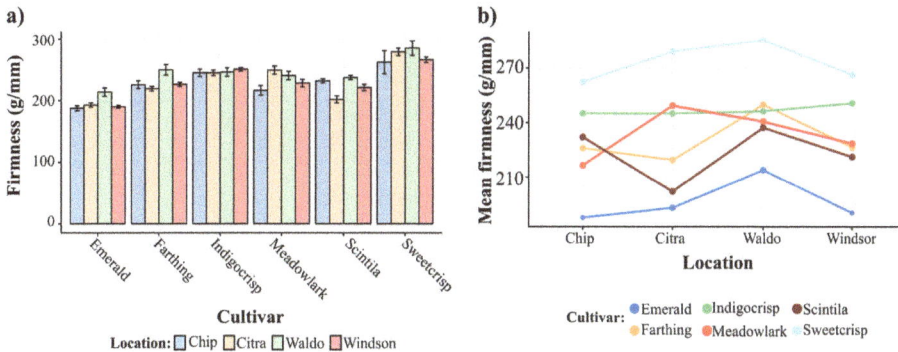

Figure 4. Genotype-by-environment interaction in SHB using the historical data recorded by the breeding program at the University of Florida. (**a**) Average of firmness for six cultivars, evaluated across four locations, during seven years; (**b**) Interaction plot showing the changes in firmness profile for six cultivars across four locations.

3.4. Genomics Tools for Fruit Firmness Improvement

Genomics-based strategies are increasingly being used to assist crop improvement. Advances in next-generation sequencing (NGS) and genotyping technologies have enabled high-throughput and relative low-cost identification of single-nucleotide polymorphisms (SNPs). By using thousands of markers spread throughout the genome, genome-wide association studies (GWAS) and genomic selection (GS) methods are becoming popular in many crop species [155]. GWAS is a methodology that investigates genetic variants in a large and genetically diverse population, testing each SNP for association to phenotypes of interest [156]; while GS uses all markers simultaneously to predict the breeding value of individuals [157]. Despite the potential of these approaches to accelerate breeding programs, there are remarkably few studies attempting to implement GWAS or GS models in blueberry.

A recent GWAS in SHB was the first attempt to identify the genetic basis of firmness variation in a breeding population [158]. SNP-firmness associations were detected for tetraploid (five associations) and diploid (three associations) gene action models only under a less stringent threshold (q-value of 0.1). Out of those, two missense variants were identified, one at a gene encoding a ubiquitin-like-specific cysteine proteinase (CUFF.36470.1) and another at a gene encoding a *S*-adenosyl-L-methionine-dependent methyltransferase (SAM-MTase) (CUFF.1480.1). Cysteine proteinases have been shown to act as post-transcriptional regulators of ripening-related proteins in tomato [159] and cysteine proteinases were also differentially expressed between firm and soft strawberry cultivars [160]. SAM-MTases catalyze transmethylation reactions in various biomolecules, acting in the biosynthesis pathway of ethylene and polyamines, which also have important roles during the ripening process and may affect fruit firmness [161–165]. Hence, some of the significant SNPs were detected within biologically plausible candidate genes affecting the trait [158]. However, individual markers explained a small portion of the phenotypic variation (less than 3.46%). These results suggested that firmness is indeed a quantitative trait, whose phenotypic expression depends on the cumulative actions of many genes with small effects and their interaction with environment. However, assuming that firmness and crispiness are correlated traits, these results might conflict with the results of Blaker (2013) [83], where crispiness segregation pattern fit the expected ratio for a monogenic trait. Ferrão et al. (2018) [158] also raised the concern that controlling for population structure in the GWAS model can strongly reduce the statistical power to detect associations when phenotypes are correlated with relationship. In this scenario, a QTL mapping would be a more suitable approach for detecting loci with large effects.

Another breakthrough in the plant breeding field is the CRISPR/Cas9 (and related variants) technologies. CRISPR/Cas9 is a molecular system composed of an endonuclease capable of precisely cutting a targeted genomic DNA by matching a pre-designed guide RNA (gRNA). Extensive research has shown that this system can be customized to inactivate, edit, or insert a gene of interest [166–168]. This technology is especially useful for monogenic traits. A successful example was the CRISPR-mediated knock-out of a canker-susceptibility gene in citrus, generating a resistant plant [169]. This technique can also be used for improving polygenic traits, however, with less evident effects [170]. To our knowledge, no study has been published applying CRISPR technologies in blueberry to date. Editing the blueberry genome for improving firmness faces two major challenges: (I) detect a candidate region with large effects on fruit firmness to be edited and (II) optimize tissue culture protocols for the species to get high editing efficiency.

We also reinforce that the absence of a high-quality reference genome for blueberry imposes additional challenges for the implementation of genomics tools. The current available genome is very fragmented, with 11,797 scaffolds and N50 of 269,026 in the 2015 version, and many predicted genes are incomplete [64]. Improving the genome contiguity by using long reads (e.g., PacBio or Nanopore sequencing) and scaffolding tools (e.g., linkage map, optical map, Hi-C, 10X Genomics Chromium, linked-reads), as well as sequencing different types of blueberries will benefit the entire blueberry research community.

4. Conclusions and Perspectives

High fruit firmness values are important for the whole blueberry industry chain. Among the main benefits, firmness is a key trait for the wide implementation of machine harvesting for fresh market, which will reduce financial and labor concerns. Considerable genetic gains have been achieved throughout the history of blueberry breeding, with modern cultivars being, on average, firmer than first-released genotypes. Despite the importance of this trait, the number of genetic and molecular studies reported in the literature is still modest. Recent insights through RNA-seq and GWAS analyses highlighted the role of cell wall degrading enzymes, cysteine proteinase, and SAM-MTase in fruit firmness variation across developmental stages and across genotypes. For the implementation of marker-assisted selection in order to accelerate the breeding process, new experimental designs are required. We highlighted the more accurate measurements of the trait by using higher number of berry samples per genotype, a broader population panels for GWAS and/or a QTL mapping approach to detect loci with large effects on firmness variation, investigate the potential of GS to predict firmness; and establish CRISPR protocols for the species. More studies in these areas have the potential to unlock an era of faster and more efficient breeding tools for this healthy and high-value fruit crop.

Supplementary Materials: The following are available online at http://www.mdpi.com/2073-4395/8/9/174/s1, Table S1: Fruit firmness survey for four cultivated blueberry types.

Author Contributions: F.C. performed the literature survey of firmness values. L.F.V.F. and J.B. performed the data analyses and interpretation. F.C., J.B., L.F.V.F., and P.M. wrote the paper. P.M. conceived and supervised writing. All authors approved the final manuscript.

Funding: This work was funded by the UF royalty fund generated by the licensing of blueberry cultivars.

Conflicts of Interest: The authors declare no conflict of interest.

References

1. Szajdek, A.; Borowska, E.J. Bioactive Compounds and Health-Promoting Properties of Berry Fruits: A Review. *Plant Foods Hum. Nutr.* **2008**, *63*, 147–156. [CrossRef] [PubMed]
2. Rodriguez-Mateos, A.; Feliciano, R.P.; Cifuentes-Gomez, T.; Spencer, J.P.E. Bioavailability of wild blueberry (poly)phenols at different levels of intake. *J. Berry Res.* **2016**, *6*, 137–148. [CrossRef]
3. Norberto, S.; Silva, S.; Meireles, M.; Faria, A.; Pintado, M.; Calhau, C. Blueberry anthocyanins in health promotion: A metabolic overview. *J. Funct. Foods* **2013**, *5*, 1518–1528. [CrossRef]
4. The Food and Agriculture Organization of the United Nations -FAOSTAT. Available online: http://www.fao.org/faostat/en/#data/QC (accessed on 17 May 2018).
5. Gallardo, R.K.; Stafne, E.T.; DeVetter, L.W.; Zhang, Q.; Li, C.; Takeda, F.; Williamson, J.; Yang, W.Q.; Cline, W.O.; Beaudry, R. Blueberry Producers' Attitudes toward Harvest Mechanization for Fresh Market. *Horttechnology* **2018**, *28*, 10–16. [CrossRef]
6. Ehlenfeldt, M.K.; Martin, R.B. A survey of fruit firmness in highbush blueberry and species-introgressed blueberry cultivars. *HortScience* **2002**, *37*, 386–389.
7. Olmstead, J.W.; Finn, C.E. Breeding highbush blueberry cultivars adapted to machine harvest for the fresh market. *Horttechnology* **2014**, *24*, 290–294.
8. Berger, C.N.; Sodha, S.V.; Shaw, R.K.; Griffin, P.M.; Pink, D.; Hand, P.; Frankel, G. Fresh fruit and vegetables as vehicles for the transmission of human pathogens. *Environ. Microbiol.* **2010**, *12*, 2385–2397. [CrossRef] [PubMed]
9. Taylor, P.; Chen, P.M.; Spotts, R.A. Changes in Ripening Behaviors of 1-MCP-Treated 'd' Anjou' Pears After Storage Changes in Ripening Behaviors of 1-MCP-Treated 'd' Anjou' Pears After Storage. *Int. J. Fruit Sci.* **2005**, *5*, 3–18. [CrossRef]
10. Moggia, C.; Graell, J.; Lara, I.; González, G.; Lobos, G.A. Firmness at Harvest Impacts Postharvest Fruit Softening and Internal Browning Development in Mechanically Damaged and Non-damaged Highbush Blueberries (*Vaccinium corymbosum* L.). *Front. Plant Sci.* **2017**, *8*. [CrossRef] [PubMed]
11. Mehra, L.K.; MacLean, D.D.; Savelle, A.T.; Scherm, H. Postharvest Disease Development on Southern Highbush Blueberry Fruit in Relation to Berry Flesh Type and Harvest Method. *Plant Dis.* **2013**, *97*, 213–221. [CrossRef]
12. Giongo, L.; Poncetta, P.; Loretti, P.; Costa, F. Texture profiling of blueberries (Vaccinium spp.) during fruit development, ripening and storage. *Postharvest Biol. Technol.* **2013**, *76*, 34–39. [CrossRef]
13. Gilbert, J.L.; Guthart, M.J.; Gezan, S.A.; Pisaroglo de Carvalho, M.; Schwieterman, M.L.; Colquhoun, T.A.; Bartoshuk, L.M.; Sims, C.A.; Clark, D.G.; Olmstead, J.W. Identifying Breeding Priorities for Blueberry Flavor Using Biochemical, Sensory, and Genotype by Environment Analyses. *PLoS ONE* **2015**, *10*, e0138494. [CrossRef] [PubMed]
14. Saftner, R.; Polashock, J.; Ehlenfeldt, M.; Vinyard, B. Instrumental and sensory quality characteristics of blueberry fruit from twelve cultivars. *Postharvest Biol. Technol.* **2008**, *49*, 19–26. [CrossRef]
15. Blaker, K.M.; Olmstead, J.W. Stone cell frequency and cell area variation of crisp and standard texture southern highbush blueberry fruit. *J. Am. Soc. Hortic. Sci.* **2014**, *139*, 553–557.
16. Forney, C.F.; Kalt, W.; Jordan, M.A.; Vinqvist-Tymchuk, M.R.; Fillmore, S.A.E. Blueberry and cranberry fruit composition during development. *J. Berry Res.* **2012**, *2*, 169–177. [CrossRef]
17. Chen, H.; Cao, S.; Fang, X.; Mu, H.; Yang, H.; Wang, X.; Xu, Q.; Gao, H. Changes in fruit firmness, cell wall composition and cell wall degrading enzymes in postharvest blueberries during storage. *Sci. Hortic. (Amsterdam)* **2015**, *188*, 44–48. [CrossRef]

18. Wang, D.; Yeats, T.H.; Uluisik, S.; Rose, J.K.C.; Seymour, G.B. Fruit Softening: Revisiting the Role of Pectin. *Trends Plant Sci.* **2018**, *23*, 302–310. [CrossRef] [PubMed]
19. Cellon, C.; Amadeu, R.R.; Olmstead, J.W.; Mattia, M.R.; Ferrao, L.F.V.; Munoz, P.R. Estimation of genetic parameters and prediction of breeding values in an autotetraploid blueberry breeding population with extensive pedigree data. *Euphytica* **2018**, *214*, 87. [CrossRef]
20. Bell, H.P.; Burchill, J. Flower development in the lowbush blueberry. *Can. J. Bot.* **1955**, *33*, 251–258. [CrossRef]
21. Gough, R.E. *The Highbush Blueberry and Its Management*; Food Products Press: Binghamton, NY, USA, 1994.
22. Chu, W.; Gao, H.; Chen, H.; Wu, W.; Fang, X. Changes in Cuticular Wax Composition of Two Blueberry Cultivars during Fruit Ripening and Postharvest Cold Storage. *J. Agric. Food Chem.* **2018**, *66*, 2870–2876. [CrossRef] [PubMed]
23. Godoy, C.; Monterubbianesi, G.; Tognetti, J. Analysis of highbush blueberry (*Vaccinium corymbosum* L.) fruit growth with exponential mixed models. *Sci. Hortic. (Amsterdam)* **2008**, *115*, 368–376. [CrossRef]
24. Jorquera-Fontena, E.; Génard, M.; Franck, N. Analysis of blueberry (*Vaccinium corymbosum* L.) fruit water dynamics during growth using an ecophysiological model. *J. Hortic. Sci. Biotechnol.* **2017**, *92*, 646–659. [CrossRef]
25. Janick, J.; Paull, R.E. *The Encyclopedia of Fruit and Nuts*; CAB International: Wallingford, UK, 2008.
26. Liu, M.; Pirrello, J.; Chervin, C.; Roustan, J.-P.; Bouzayen, M. Ethylene control of fruit ripening: Revisiting the complex network of transcriptional regulation. *Plant Physiol.* **2015**, *169*, 2380–2390. [CrossRef] [PubMed]
27. Kumar, R.; Khurana, A.; Sharma, A.K. Role of plant hormones and their interplay in development and ripening of fleshy fruits. *J. Exp. Bot.* **2013**, *65*, 4561–4575. [CrossRef] [PubMed]
28. Haji, T.; Yaegaki, H.; Yamaguchi, M. Softening of stony hard peach by ethylene and the induction of endogenous ethylene by 1-aminocyclopropane-1-carboxylic acid (ACC). *J. Jpn. Soc. Hortic. Sci.* **2003**, *72*, 212–217. [CrossRef]
29. Hoeberichts, F.A.; Van Der Plas, L.H.W.; Woltering, E.J. Ethylene perception is required for the expression of tomato ripening-related genes and associated physiological changes even at advanced stages of ripening. *Postharvest Biol. Technol.* **2002**, *26*, 125–133. [CrossRef]
30. Suzuki, A.; Kikuchi, T.; Aoba, K. Changes of ethylene evolution, ACC content, ethylene forming enzyme activity and respiration in fruits of highbush blueberry. *J. Jpn. Soc. Hortic. Sci.* **1997**, *66*, 23–27. [CrossRef]
31. Ban, T.; Kugishima, M.; Ogata, T.; Shiozaki, S.; Horiuchi, S.; Ueda, H. Effect of ethephon (2-chloroethylphosphonic acid) on the fruit ripening characters of rabbiteye blueberry. *Sci. Hortic. (Amsterdam)* **2007**, *112*, 278–281. [CrossRef]
32. Costa, D.V.T.A.; Pintado, M.; Almeida, D.P.F. Postharvest ethylene application affects anthocyanin content and antioxidant activity of blueberry cultivars. *ISHS Acta Hortic.* **2014**, *1017*, 525–530. [CrossRef]
33. Chiabrando, V.; Giacalone, G. Shelf-life extension of highbush blueberry using 1-methylcyclopropene stored under air and controlled atmosphere. *Food Chem.* **2011**, *126*, 1812–1816. [CrossRef] [PubMed]
34. Wang, S.; Zhou, Q.; Zhou, X.; Wei, B.; Ji, S. The effect of ethylene absorbent treatment on the softening of blueberry fruit. *Food Chem.* **2018**, *246*, 286–294. [CrossRef] [PubMed]
35. DeLong, J.M.; Prange, R.K.; Bishop, C.; Harrison, P.A.; Ryan, D.A.J. The influence of 1-MCP on shelf-life quality of highbush blueberry. *HortScience* **2003**, *38*, 417–418.
36. Deng, J.; Shi, Z.; Li, X.; Liu, H. Effects of cold storage and 1-methylcyclopropene treatments on ripening and cell wall degrading in rabbiteye blueberry (Vaccinium ashei) fruit. *Food Sci. Technol. Int.* **2014**, *20*, 287–298. [CrossRef] [PubMed]
37. Blaker, K.M.; Olmstead, J.W. Effects of Preharvest Applications of 1-Methylcyclopropene on Fruit Firmness in Southern Highbush Blueberry. *ISHS Acta Hortic.* **2014**, *1017*, 71–75. [CrossRef]
38. MacLean, D.D.; NeSmith, D.S. Rabbiteye blueberry postharvest fruit quality and stimulation of ethylene production by 1-methylcyclopropene. *HortScience* **2011**, *46*, 1278–1281.
39. Sun, Y.; Hou, Z.; Su, S.; Yuan, J. Effects of ABA, GA3 and NAA on fruit development and anthocyanin accumulation in blueberry. *J. South China Agric. Univ.* **2013**, *34*, 6–11.
40. Buran, T.J.; Sandhu, A.K.; Azeredo, A.M.; Bent, A.H.; Williamson, J.G.; Gu, L. Effects of exogenous abscisic acid on fruit quality, antioxidant capacities, and phytochemical contents of southern high bush blueberries. *Food Chem.* **2012**, *132*, 1375–1381. [CrossRef] [PubMed]
41. Brummell, D.A. Cell wall disassembly in ripening fruit. *Funct. Plant Biol.* **2006**, *33*, 103. [CrossRef]

42. Cell, P.; Sørensen, I.; Willats, W.G.T. *The Plant Cell Wall*; Humana Press: Totowa, NJ, USA, 2011; Volume 715, pp. 115–121. ISBN 978-1-61779-007-2.

43. Daher, F.B.; Braybrook, S.A. How to let go: Pectin and plant cell adhesion. *Front. Plant Sci.* **2015**, *6*. [CrossRef] [PubMed]

44. Carpita, N.C.; Gibeaut, D.M. Structural models of primary cell walls in flowering plants: Consistency of molecular structure with the physical properties of the walls during growth. *Plant J.* **1993**, *3*, 1–30. [CrossRef] [PubMed]

45. Huber, D.J. Polyuronide degradation and hemicellulose modifications in ripening tomato fruit. *J. Am. Soc. Hortic. Sci.* **1983**, *108*, 405–409.

46. Cutillas-Iturralde, A.; Zarra, I.; Fry, S.C.; Lorences, E.P. Implication of persimmon fruit hemicellulose metabolism in the softening process. Importance of xyloglucan endotransglycosylase. *Physiol. Plant.* **1994**, *91*, 169–176. [CrossRef]

47. Brummell, D.A.; Schröder, R. Xylan metabolism in primary cell walls. *N. Z. J. For. Sci.* **2009**, *39*, 125–143.

48. Atkinson, R.G.; Sutherland, P.W.; Johnston, S.L.; Gunaseelan, K.; Hallett, I.C.; Mitra, D.; Brummell, D.A.; Schröder, R.; Johnston, J.W.; Schaffer, R.J. Down-regulation of POLYGALACTURONASE1 alters firmness, tensile strength and water loss in apple (Malus x domestica) fruit. *BMC Plant Biol.* **2012**, *12*, 129. [CrossRef] [PubMed]

49. Vicente, A.R.; Ortugno, C.; Rosli, H.; Powell, A.L.T.; Greve, L.C.; Labavitch, J.M. Temporal Sequence of Cell Wall Disassembly Events in Developing Fruits. 2. Analysis of Blueberry (Vaccinium Species). *J. Agric. Food Chem.* **2007**, *55*, 4125–4130. [CrossRef] [PubMed]

50. Beaudry, R.; Hanson, E.J.; Beggs, J.L.; Beaudry, R.M. Applying Calcium Chloride Postharvest to Improve Highbush Blueberry Firmness Applying Calcium Chloride Postharvest to Improve Highbush Blueberry Firmness. *HortScience* **2016**, *28*, 2–4.

51. Angeletti, P.; Castagnasso, H.; Miceli, E.; Terminiello, L.; Concellón, A.; Chaves, A.; Vicente, A.R. Effect of preharvest calcium applications on postharvest quality, softening and cell wall degradation of two blueberry (Vaccinium corymbosum) varieties. *Postharvest Biol. Technol.* **2010**, *58*, 98–103. [CrossRef]

52. Buchanan, B.B.; Gruissem, W.; Jones, R.L. *Biochemistry & Molecular Biology of Plants*; American Society of Plant Physiologists: Rockville, MD, USA, 2000; Volume 40.

53. Lara, I.; García, P.; Vendrell, M. Modifications in cell wall composition after cold storage of calcium-treated strawberry (Fragaria× ananassa Duch.) fruit. *Postharvest Biol. Technol.* **2004**, *34*, 331–339. [CrossRef]

54. Conway, W.S.; Sams, C.E. The effects of postharvest infiltration of calcium, magnesium, or strontium on decay, firmness, respiration, and ethylene production in apples. *J. Am. Soc. Hortic. Sci.* **1987**, *112*, 300–303.

55. Conway, W.S. Possible Mechanisms by Which Postharvest Calcium Treatment Reduces Decay in Apples. *Phytopathology* **1984**, *74*, 208. [CrossRef]

56. Sams, C.E.; Conway, W.S.; Abbott, J.A.; Lewis, R.J.; Ben-Shalom, N. Firmness and decay of apples following postharvest pressure infiltration of calcium and heat treatment. *J. Am. Soc. Hortic. Sci.* **1993**, *118*, 623–627.

57. Al-Banna, M.K.S.; Jinks, J.L. Indirect selection for environmental sensitivity in Nicotiana rustica. *Hered. (Edinb)* **1984**, *52*, 297–301. [CrossRef]

58. Bonomelli, C.; Ruiz, R. Effects of foliar and soil calcium application on yield and quality of table grape cv.'Thompson Seedless'. *J. Plant Nutr.* **2010**, *33*, 299–314. [CrossRef]

59. Ciccarese, A.; Stellacci, A.M.; Gentilesco, G.; Rubino, P. Effectiveness of pre-and post-veraison calcium applications to control decay and maintain table grape fruit quality during storage. *Postharvest Biol. Technol.* **2013**, *75*, 135–141. [CrossRef]

60. García, J.M.; Herrera, S.; Morilla, A. Effects of Postharvest Dips in Calcium Chloride on Strawberry. *J. Agric. Food Chem.* **1996**, *44*, 30–33. [CrossRef]

61. Hernández-Muñoz, P.; Almenar, E.; Ocio, M.J.; Gavara, R. Effect of calcium dips and chitosan coatings on postharvest life of strawberries (Fragaria x ananassa). *Postharvest Biol. Technol.* **2006**, *39*, 247–253. [CrossRef]

62. Hernández-Muñoz, P.; Almenar, E.; Del Valle, V.; Velez, D.; Gavara, R. Effect of chitosan coating combined with postharvest calcium treatment on strawberry (Fragaria x ananassa) quality during refrigerated storage. *Food Chem.* **2008**, *110*, 428–435. [CrossRef] [PubMed]

63. Lombard, V.; Golaconda Ramulu, H.; Drula, E.; Coutinho, P.M.; Henrissat, B. The carbohydrate-active enzymes database (CAZy) in 2013. *Nucleic Acids Res.* **2014**, *42*, 490–495. [CrossRef] [PubMed]

64. Gupta, V.; Estrada, A.D.; Blakley, I.; Reid, R.; Patel, K.; Meyer, M.D.; Andersen, S.U.; Brown, A.F.; Lila, M.A.; Loraine, A.E. RNA-Seq analysis and annotation of a draft blueberry genome assembly identifies candidate genes involved in fruit ripening, biosynthesis of bioactive compounds, and stage-specific alternative splicing. *Gigascience* **2015**, *4*, 5. [CrossRef] [PubMed]

65. Cantarel, B.L.; Coutinho, P.M.; Rancurel, C.; Bernard, T.; Lombard, V.; Henrissat, B. The Carbohydrate-Active EnZymes database (CAZy): An expert resource for glycogenomics. *Nucleic Acids Res.* **2008**, *37* (Suppl. 1), D233–D238. [CrossRef] [PubMed]

66. Yin, Y.; Mao, X.; Yang, J.; Chen, X.; Mao, F.; Xu, Y. dbCAN: A web resource for automated carbohydrate-active enzyme annotation. *Nucleic Acids Res.* **2012**, *40*, W445–W451. [CrossRef] [PubMed]

67. Asif, M.H.; Lakhwani, D.; Pathak, S.; Gupta, P.; Bag, S.K.; Nath, P.; Trivedi, P.K. Transcriptome analysis of ripe and unripe fruit tissue of banana identifies major metabolic networks involved in fruit ripening process. *BMC Plant Biol.* **2014**, *14*, 316. [CrossRef] [PubMed]

68. Rose, J.K.C.; Bennett, A.B. Cooperative disassembly of the cellulose–xyloglucan network of plant cell walls: Parallels between cell expansion and fruit ripening. *Trends Plant Sci.* **1999**, *4*, 176–183. [CrossRef]

69. Goulao, L.F.; Oliveira, C.M. Cell wall modifications during fruit ripening: When a fruit is not the fruit. *Trends Food Sci. Technol.* **2008**, *19*, 4–25. [CrossRef]

70. Chandrasekar, B.; van der Hoorn, R.A.L. Beta galactosidases in Arabidopsis and tomato—A mini review. *Biochem. Soc. Trans.* **2016**, *44*, 150–158. [CrossRef] [PubMed]

71. Hossain, M.A.; Rana, M.M.; Kimura, Y.; Roslan, H.A. Changes in Biochemical Characteristics and Activities of Ripening Associated Enzymes in Mango Fruit during the Storage at Different Temperatures. *Biomed Res. Int.* **2014**, *2014*, 1–11. [CrossRef] [PubMed]

72. Uluisik, S.; Chapman, N.H.; Smith, R.; Poole, M.; Adams, G.; Gillis, R.B.; Besong, T.M.D.; Sheldon, J.; Stiegelmeyer, S.; Perez, L.; et al. Genetic improvement of tomato by targeted control of fruit softening. *Nat. Biotechnol.* **2016**, *34*, 950–952. [CrossRef] [PubMed]

73. Wu, B.; Gao, L.; Gao, J.; Xu, Y.; Liu, H.; Cao, X.; Zhang, B.; Chen, K. Genome-Wide Identification, Expression Patterns, and Functional Analysis of UDP Glycosyltransferase Family in Peach (*Prunus persica* L. Batsch). *Front. Plant Sci.* **2017**, *8*. [CrossRef] [PubMed]

74. Carbone, F.; Preuss, A.; De Vos, R.C.H.; D'Amico, E.; Perrotta, G.; Bovy, A.G.; Martens, S.; Rosati, C. Developmental, genetic and environmental factors affect the expression of flavonoid genes, enzymes and metabolites in strawberry fruits. *Plant. Cell Environ.* **2009**, *32*, 1117–1131. [CrossRef] [PubMed]

75. Paniagua, A.C.; East, A.R.; Hindmarsh, J.P.; Heyes, J.A. Moisture loss is the major cause of firmness change during postharvest storage of blueberry. *Postharvest Biol. Technol.* **2013**, *79*, 13–19. [CrossRef]

76. Moggia, C.; Beaudry, R.M.; Retamales, J.B.; Lobos, G.A. Variation in the impact of stem scar and cuticle on water loss in highbush blueberry fruit argue for the use of water permeance as a selection criterion in breeding. *Postharvest Biol. Technol.* **2017**, *132*, 88–96. [CrossRef]

77. Moggia, C.; Graell, J.; Lara, I.; Schmeda-Hirschmann, G.; Thomas-Valdés, S.; Lobos, G.A. Fruit characteristics and cuticle triterpenes as related to postharvest quality of highbush blueberries. *Sci. Hortic. (Amsterdam)*. **2016**, *211*, 449–457. [CrossRef]

78. Sun, X.; Narciso, J.; Wang, Z.; Ference, C.; Bai, J.; Zhou, K. Effects of Chitosan-Essential Oil Coatings on Safety and Quality of Fresh Blueberries. *J. Food Sci.* **2014**, *79*. [CrossRef] [PubMed]

79. Abugoch, L.; Tapia, C.; Plasencia, D.; Pastor, A.; Castro-Mandujano, O.; López, L.; Escalona, V.H. Shelf-life of fresh blueberries coated with quinoa protein/chitosan/sunflower oil edible film. *J. Sci. Food Agric.* **2016**, *96*, 619–626. [CrossRef] [PubMed]

80. Harker, F.R.; Redgwell, R.J.; Hallett, I.C.; Murray, S.H.; Carter, G. Chapter 2: Texture of fresh fruit. In *Horticultural Reviews*; Janick, J., Ed.; Wiley Online Library: Hoboken, NJ, USA, 1997; Volume 20, pp. 121–224.

81. Allan-Wojtas, P.M.; Forney, C.F.; Carbyn, S.E.; Nicholas, K.U.K.G. Microstructural Indicators of Quality-related Characteristics of Blueberries—An Integrated Approach. *LWT—Food Sci. Technol.* **2001**, *34*, 23–32. [CrossRef]

82. Fava, J.; Alzamora, S.M.; Castro, M.A. Structure and Nanostructure of the Outer Tangential Epidermal Cell Wall in *Vaccinium corymbosum* L. (Blueberry) Fruits by Blanching, Freezing-Thawing and Ultrasound. *Food Sci. Technol. Int.* **2006**, *12*, 241–251. [CrossRef]

83. Blaker, K.M. *Comparison of Crisp and Standard Fruit Texture in Southern Highbush Blueberry Using Instrumental and Sensory Panel Techniques*; University of Florida: Gainesville, FL, USA, 2013; ISBN 1303820617.

84. Abbott, J.; Watada, A.E.; Massie, D.R. Effe-gi, Magness-Taylor, and Instron fruit pressure testing devices for apples, peaches, and nectarines. *J. Am. Soc. Hortic. Sci.* **1976**, *101*, 698–700.

85. Mason, H.; Mason, A. Apple Tree Named "Rosy Glow". U.S. Patent 10/712,783, 26 August 2004.

86. Braun, T. Apple tree named 'Fuji Fubrax'. U.S. Patent 11/355,401, 29 April 2008.

87. Schmider, E.; Braun, T. Apple tree named Golden Parsi. U.S. Patent 12/798,834, 13 October 2011.

88. Maillard, A.; Maillard, L. Apple tree named 'REGALSTAR'. U.S. Patent 13/999,809, 31 May 2016.

89. Harker, F.R.; Maindonald, J.H.; Jackson, P.J. Penetrometer Measurement of Apple and Kiwifruit Firmness: Operator and Instrument Differences. *J. Am. Soc. Hortic. Sci.* **1996**, *121*, 927–936.

90. Lehman-Salada, L. Instrument and operator effects on apple firmness readings. *HortScience* **1996**, *31*, 994–997.

91. De Belie, N.; Schotte, S.; Coucke, P.; De Baerdemaeker, J. Development of an automated monitoring device to quantify changes in firmness of apples during storage. *Postharvest Biol. Technol.* **2000**, *18*, 1–8. [CrossRef]

92. Plocharski, W.; Konopacka, D.; Zwierz, J. Comparison of Magness-Taylor's pressure test with mechanical, non-destructive methods of apple and pear firmness measurements. *Int. Agrophys.* **2000**, *14*, 311–318.

93. Abbott, J.A.; Liljedahl, L.A. Relationship of sonic resonant frequency to compression tests and Magness-Taylor firmness of apples during refrigerated storage. *Trans. ASAE* **1994**, *37*, 1211–1215. [CrossRef]

94. Abbott, J.A.; Massie, D.R.; Upchurch, B.L.; Hruschka, W.R. Nondestructive sonic firmness measurement of apples. *Trans. ASAE* **1995**, *38*, 1461–1466. [CrossRef]

95. Lu, R.; Peng, Y. Hyperspectral scattering for assessing peach fruit firmness. *Biosyst. Eng.* **2006**, *93*, 161–171. [CrossRef]

96. Abbott, J.A. Firmness Measurement of Freshly Harvested 'Delicious' Apples by Sensory Methods, Sonic Transmission, Magness-Taylor, and Compression. *J. Am. Soc. Hortic. Sci.* **1994**, *119*, 510–515.

97. Elmasry, G.; Wang, N.; Vigneault, C. Postharvest Biology and Technology Detecting chilling injury in Red Delicious apple using hyperspectral imaging and neural networks. *Postharvest Biol. Technol.* **2009**, *52*, 1–8. [CrossRef]

98. Lu, R. Multispectral imaging for predicting firmness and soluble solids content of apple fruit. *Postharvest Biol. Technol.* **2004**, *31*, 147–157. [CrossRef]

99. Finn, C.E.; Luby, J.J. Inheritance of fruit quality traits in blueberry. *J. Am. Soc. Hortic. Sci.* **1992**, *117*, 617–621.

100. Ali, S.; Zaman, Q.U.; Schumann, A.W.; Udenigwe, C.C.; Farooque, A.A. Impact of fruit ripening parameters on harvesting efficiency of the wild blueberry harvester. In *ASABE Annual International Meeting*; American Society of Agricultural and Biological Engineers: St. Joseph, MI, USA, 2016.

101. Vilela, A.; Gonçalves, B.; Ribeiro, C.; Fonseca, A.T.; Correia, S.; Fernandes, H.; Ferreira, S.; Bacelar, E.; Silva, A.P. Study of Textural, Chemical, Color and Sensory Properties of Organic Blueberries Harvested in Two Distinct Years: A Chemometric Approach. *J. Texture Stud.* **2016**, *47*, 199–207. [CrossRef]

102. Concha-Meyer, A.; Eifert, J.D.; Williams, R.C.; Marcy, J.E.; Welbaum, G.E. Shelf life determination of fresh blueberries (Vaccinium corymbosum) stored under controlled atmosphere and ozone. *Int. J. food Sci.* **2015**. [CrossRef] [PubMed]

103. Hu, M.-H.; Dong, Q.-L.; Liu, B.-L.; Opara, U.L. Prediction of mechanical properties of blueberry using hyperspectral interactance imaging. *Postharvest Biol. Technol.* **2016**, *115*, 122–131. [CrossRef]

104. Brazelton, D.M.; Wagner, A.L. Blueberry Plant Named 'Last Call'. 2015. Available online: https://patents.google.com/patent/USPP25386P3 (accessed on 15 August 2018).

105. NeSmith, D.S. 'Suziblue'Southern Highbush Blueberry. *HortScience* **2010**, *45*, 142–143.

106. Grajkowski, J.; Ochman, I.; Muliński, Z. Firmness and antioxidant capacity of highbush blueberry (*Vaccinium corymbosum* L.) grown on three types of organic bed. *Veg. Crop. Res. Bull.* **2007**, *66*, 155–159. [CrossRef]

107. Blaker, K.M.; Plotto, A.; Baldwin, E.A.; Olmstead, J.W. Correlation between sensory and instrumental measurements of standard and crisp-texture southern highbush blueberries (*Vaccinium corymbosum* L. interspecific hybrids). *J. Sci. Food Agric.* **2014**, *94*, 2785–2793. [CrossRef] [PubMed]

108. NeSmith, D.S.; Draper, A.D.; Spiers, J.M. Palmetto'Southern Highbush Blueberry. *HortScience* **2004**, *39*, 1774–1775.

109. Pavlis, G.C. Blueberry fruit quality and yield as affected by fertilization. *ISHS Acta Hortic.* **2004**, 353–356. [CrossRef]

110. Ochmian, I.; Grajkowski, J.; Skupień, K. Effect of substrate type on the field performance and chemical composition of highbush blueberry cv. Patriot. *Agric. Food Sci.* **2010**, *19*, 69–80. [CrossRef]

111. Retamales, J.B.; Lobos, G.A.; Romero, S.; Godoy, R.; Moggia, C. Repeated applications of CPPU on highbush blueberry cv. Duke increase yield and enhance fruit quality at harvest and during postharvest. *Chil. J. Agric. Res.* **2014**, *74*, 157–161. [CrossRef]

112. Yu, P.; Li, C.; Takeda, F.; Krewer, G. Visual bruise assessment and analysis of mechanical impact measurement in southern highbush blueberries. *Appl. Eng. Agric.* **2014**, *30*, 29–37. [CrossRef]

113. Stringer, S.J.; Draper, A.D.; Spiers, J.M.; Marshall, D.A.; Smith, B.J. 'Pearl'Southern Highbush Blueberry. *HortScience* **2013**, *48*, 130–131.

114. Lobos, G.A.; Callow, P.; Hancock, J.F. The effect of delaying harvest date on fruit quality and storage of late highbush blueberry cultivars (*Vaccinium corymbosum* L.). *Postharvest Biol. Technol.* **2014**, *87*, 133–139. [CrossRef]

115. Ehret, D.L.; Frey, B.; Forge, T.; Helmer, T.; Bryla, D.R. Effects of drip irrigation configuration and rate on yield and fruit quality of young highbush blueberry plants. *HortScience* **2012**, *47*, 414–421.

116. Hicklenton, P.; Forney, C.; Domytrak, C. Use of row covers and post harvest storage techniques to alter maturity and marketing period for highbush blueberries. *ISHS Acta Hortic.* **2002**, 287–295. [CrossRef]

117. Sargent, S.A.; Berry, A.D.; Brecht, J.K.; Santana, M.; Zhang, S.; Ristow, N. Studies on quality of southern highbush blueberry cultivars: Effects of pulp temperature, impact and hydrocooling. *ISHS Acta Hortic.* **2017**, *1180*, 497–502. [CrossRef]

118. Brazelton, D.M.; Wagner, A.L. Blueberry Plant named'CLOCKWORK'. 2013. Available online: https://patents.google.com/patent/US20130239264P1 (accessed on 15 August 2018).

119. Yang, W.Q.; Andrews, H.E.; Basey, A. Blueberry rootstock: Selection, evaluation, and field performance of grafted blueberry plants. *ISHS Acta Hortic.* **2016**, *1117*, 119–124. [CrossRef]

120. Brazelton, D.M.; Wagner, A.L. Blueberry Plant Named 'Cargo' 2014. Available online: https://patents.google.com/patent/US20130239260P1 (accessed on 15 August 2018).

121. Brazelton, D.M.; Wagner, A.L. Blueberry Plant Named 'Blue Ribbon' 2014. Available online: https://patents.google.com/patent/US20130239265P1 (accessed on 15 August 2018).

122. Brazelton, D.M.; Wagner, A.L. Blueberry Plant Named 'Top Shelf' 2014. Available online: https://patents.google.com/patent/US20130239261P1 (accessed on 15 August 2018).

123. Moggia, C.; González, C.; Lobos, G.A.; Bravo, C.; Valdés, M.; Lara, I.; Graell, J. Changes in quality and maturity of 'Duke' and 'Brigitta' blueberries during fruit development: Postharvest implications. *Acta Hortic.* **2018**, *1194*, 1495–1501. [CrossRef]

124. Patel, N. Blueberry Plant Named 'Hortblue Poppins' 2011. Available online: https://patents.google.com/patent/USPP21881P3 (accessed on 15 August 2018).

125. Ballington, J.R.; Bland, W.T. Blueberry Plant Named 'Heintooga' 2017. Available online: https://patents.google.com/patent/US20170188494P1 (accessed on 15 August 2018).

126. Strik, B.C.; Vance, A.J.; Finn, C.E. Northern Highbush Blueberry Cultivars Differed in Yield and Fruit Quality in Two Organic Production Systems from Planting to Maturity. *HortScience* **2017**, *52*, 844–851. [CrossRef]

127. Rodríguez-Armenta, H.P.; Olmstead, J.W.; Lyrene, P.M. Characterization of backcross blueberry populations created to introgress Vaccinium arboreum traits into southern highbush blueberry. *ISHS Acta Hortic.* **2016**, *1180*, 435–444. [CrossRef]

128. Ballington, J.; Rooks, S. Blueberry Named 'Carteret' 2007. Available online: https://patents.google.com/patent/US20070143882 (accessed on 15 August 2018).

129. Ballington, J.; Rooks, S. Blueberry named 'New Hanover' 2007. Available online: https://patents.google.com/patent/USPP19990P3 (accessed on 15 August 2018).

130. Ballington, J.R.; Rooks, S.D. Blueberry named 'Robeson' 2009. Available online: https://patents.google.com/patent/USPP19756P3 (accessed on 15 August 2018).

131. Almutairi, K.; Bryla, D.R.; Strik, B.C. Potential of Deficit Irrigation, Irrigation Cutoffs, and Crop Thinning to Maintain Yield and Fruit Quality with Less Water in Northern Highbush Blueberry. *HortScience* **2017**, *52*, 625–633. [CrossRef]

132. Strik, B.; Buller, G. Nitrogen fertilization rate, sawdust mulch, and pre-plant incorporation of sawdust-term impact on yield, fruit quality, and soil and plant nutrition in 'elliott'. *ISHS Acta Hortic.* **2014**, *1017*, 269–275. [CrossRef]

133. NeSmith, D.S.; Prussia, S.; Tetteh, M.; Krewer, G. Firmness losses of rabbiteye blueberries (Vaccinium ashei Reade) during harvesting and handling. *ISHS Acta Hortic.* **2000**, 287–293. [CrossRef]

134. Yang, W.Q.; Harpole, J.; Finn, C.E.; Strik, B.C. Evaluating berry firmness and total soluble solids of newly released highbush blueberry cultivars. *ISHS Acta Hortic.* **2008**, 863–868. [CrossRef]
135. Li, C.; Luo, J.; MacLean, D. A novel instrument to delineate varietal and harvest effects on blueberry fruit texture during storage. *J. Sci. Food Agric.* **2011**, *91*, 1653–1658. [CrossRef] [PubMed]
136. NeSmith, D.S.; Draper, A.D. 'Camellia'southern highbush blueberry. *J. Am. Pomol. Soc.* **2007**, *61*, 34–37.
137. NeSmith, D.S.; Nunez-Barrios, A.; Prussia, S.E.; Aggarwal, D. Postharvest berry quality of six rabbiteye blueberry cultivars in response to temperature. *J. Am. Pomol. Soc.* **2005**, *59*, 13–17.
138. Døving, A.; Måge, F.; Vestrheim, S. Methods for Testing Strawberry Fruit Firmness Methods for Testing Strawberry Fruit Firmness: A Review. *Small Fruits Rev.* **2008**, *8851*, 37–41. [CrossRef]
139. Ballington, J.R.; Ballinger, W.E.; Mainland, C.M.; Swallow, W.H.; Maness, E.P. Ripening period of Vaccinium species in southeastern North Carolina [Blueberry, breeding for both early-and late-ripening Vaccinium genotypes]. *J. Am. Soc. Hortic. Sci.* **1984**, *109*, 392–396.
140. Boches, P.; Bassil, N.V.; Rowland, L. Genetic diversity in the highbush blueberry evaluated with microsatellite markers. *J. Am. Soc. Hortic. Sci.* **2006**, *131*, 674–686.
141. Hancock, J.F.; Lyrene, P.; Finn, C.E.; Vorsa, N.; Lobos, G.A. Blueberries and cranberries. In *Temperate Fruit Crop Breeding*; James, F.H., Ed.; Springer: Berlin, Germany, 2008; pp. 115–150.
142. Goldy, R.G.; Lyrene, P.M. In vitro colchicine treatment of 4× blueberries, Vaccinium sp. [Polyploid induction, genotype effect]. *J. Am. Soc. Hortic. Sci.* **1984**, *109*, 336–338.
143. Chavez, D.J.; Lyrene, P.M. Effects of self-pollination and cross-pollination of Vaccinium darrowii (Ericaceae) and other low-chill blueberries. *HortScience* **2009**, *44*, 1538–1541.
144. Ehlenfeldt, M.K.; Draper, A.D.; Clark, J.R. Performance of southern highbush blueberry cultivars released by the US Department of Agriculture and cooperating state agricultural experiment stations. *Horttechnology* **1995**, *5*, 127–130.
145. Coville, F.V. Improving the wild blueberry. In *Yearbook of Agriculture*; U.S. Department of Agriculture; U.S. Government Printing Office: Washington, DC, USA, 1937; pp. 559–574.
146. Sharpe, R.H.; Sherman, W.B. "Floridablue and Sharpblue": Two new blueberries for central Florida. *Circ. Fla Coop Ext. Serv. Inst. Food Agric. Sci. Univ. Fla* **1976**, *240*, 6.
147. Fillion, L.; Kilcast, D. Consumer perception of crispness and crunchiness in fruits and vegetables. *Food Qual. Prefer.* **2002**, *13*, 23–29. [CrossRef]
148. Aalders, L.E.; Hall, I.V. A study of variation in fruit yield and related characters in two diallels of the lowbush blueberry, Vaccinium angustifolium Ait. *Can. J. Genet. Cytol.* **1975**, *17*, 401–404. [CrossRef]
149. Erb, W.A.; Draper, A.D.; Galletta, G.J.; Swartz, H.J. Combining ability for plant and fruit traits of interspecific blueberry progenies on mineral soil. *J. Am. Soc. Hortic. Sci.* **1990**, *115*, 1025–1028.
150. Scalzo, J.; Sguigna, V.; Mezzetti, B.; Stanley, J.; Alspach, P. Variation of fruit traits in highbush blueberry seedlings from a factorial cross. *ISHS Acta Hortic.* **2010**, 79–83. [CrossRef]
151. Scalzo, J.; Stanley, J.; Alspach, P.; Mezzetti, B. Preliminary evaluation of fruit traits and phytochemicals in a highbush blueberry seedling population. *J. Berry Res.* **2013**, *3*, 103–111. [CrossRef]
152. Connor, A.M.; Luby, J.J.; Tong, C.B.S.; Finn, C.E.; Hancock, J.F. Genotypic and environmental variation in antioxidant activity, total phenolic content, and anthocyanin content among blueberry cultivars. *J. Am. Soc. Hortic. Sci.* **2002**, *127*, 89–97.
153. Finn, C.E.; Hancock, J.F.; Mackey, T.; Serce, S. Genotype× environment interactions in highbush blueberry (Vaccinium sp. L.) families grown in Michigan and Oregon. *J. Am. Soc. Hortic. Sci.* **2003**, *128*, 196–200.
154. Isik, F.; Holland, J.; Maltecca, C. *Genetic Data Analysis for Plant and Animal Breeding*; Springer: New York, NY, USA, 2017; Volume 1, ISBN 3319551779.
155. Hamblin, M.T.; Buckler, E.S.; Jannink, J.-L. Population genetics of genomics-based crop improvement methods. *Trends Genet.* **2011**, *27*, 98–106. [CrossRef] [PubMed]
156. Visscher, P.M.; Brown, M.A.; McCarthy, M.I.; Yang, J. Five years of GWAS discovery. *Am. J. Hum. Genet.* **2012**, *90*, 7–24. [CrossRef] [PubMed]
157. Jannink, J.-L.; Lorenz, A.J.; Iwata, H. Genomic selection in plant breeding: From theory to practice. *Brief. Funct. Genom.* **2010**, *9*, 166–177. [CrossRef] [PubMed]
158. Ferrão, L.F.V.; Benevenuto, J.; de Bem Oliveira, I.; Cellon, C.; Olmstead, J.; Kirst, M.; Resende, M.F.R.; Munoz, P. Insights into the genetic basis of blueberry fruit-related traits using diploid and polyploid models in a GWAS context. *Front. Ecol. Evol.* **2018**, *6*, 107. [CrossRef]

159. Wang, W.; Cai, J.; Wang, P.; Tian, S.; Qin, G. Post-transcriptional regulation of fruit ripening and disease resistance in tomato by the vacuolar protease SlVPE3. *Genom. Biol.* **2017**, *18*, 47. [CrossRef] [PubMed]

160. Salentijn, E.M.J.; Aharoni, A.; Schaart, J.G.; Boone, M.J.; Krens, F.A. Differential gene expression analysis of strawberry cultivars that differ in fruit-firmness. *Physiol. Plant.* **2003**, *118*, 571–578. [CrossRef]

161. Roje, S. S-Adenosyl-l-methionine: Beyond the universal methyl group donor. *Phytochemistry* **2006**, *67*, 1686–1698. [CrossRef] [PubMed]

162. Paul, V.; Pandey, R.; Srivastava, G.C. The fading distinctions between classical patterns of ripening in climacteric and non-climacteric fruit and the ubiquity of ethylene—An overview. *J. Food Sci. Technol.* **2012**, *49*, 1–21. [CrossRef] [PubMed]

163. Van de Poel, B.; Bulens, I.; Oppermann, Y.; Hertog, M.L.A.T.M.; Nicolai, B.M.; Sauter, M.; Geeraerd, A.H. S-adenosyl-L-methionine usage during climacteric ripening of tomato in relation to ethylene and polyamine biosynthesis and transmethylation capacity. *Physiol. Plant.* **2013**, *148*, 176–188. [CrossRef] [PubMed]

164. Singh, R.; Rastogi, S.; Dwivedi, U.N. Phenylpropanoid Metabolism in Ripening Fruits. *Compr. Rev. Food Sci. Food Saf.* **2010**, *9*, 398–416. [CrossRef]

165. Moffatt, B.A.; Weretilnyk, E.A. Sustaining S-adenosyl-methionine-dependent methyltransferase activity in plant cells. *Physiol. Plant.* **2001**, *113*, 435–442. [CrossRef]

166. Tang, X.; Lowder, L.G.; Zhang, T.; Malzahn, A.A.; Zheng, X.; Voytas, D.F.; Zhong, Z.; Chen, Y.; Ren, Q.; Li, Q. A CRISPR–Cpf1 system for efficient genome editing and transcriptional repression in plants. *Nat. Plants* **2017**, *3*, 17018. [CrossRef] [PubMed]

167. Puchta, H. Applying CRISPR/Cas for genome engineering in plants: The best is yet to come. *Curr. Opin. Plant Biol.* **2017**, *36*, 1–8. [CrossRef] [PubMed]

168. Jaganathan, D.; Ramasamy, K.; Sellamuthu, G.; Jayabalan, S.; Venkataraman, G. CRISPR for Crop Improvement: An Update Review. *Front. Plant Sci.* **2018**, *9*. [CrossRef] [PubMed]

169. Peng, A.; Chen, S.; Lei, T.; Xu, L.; He, Y.; Wu, L.; Yao, L.; Zou, X. Engineering canker-resistant plants through CRISPR/Cas9-targeted editing of the susceptibility gene CsLOB1 promoter in citrus. *Plant Biotechnol. J.* **2017**, *15*, 1509–1519. [CrossRef] [PubMed]

170. Shi, J.; Gao, H.; Wang, H.; Lafitte, H.R.; Archibald, R.L.; Yang, M.; Hakimi, S.M.; Mo, H.; Habben, J.E. ARGOS8 variants generated by CRISPR-Cas9 improve maize grain yield under field drought stress conditions. *Plant Biotechnol. J.* **2017**, *15*, 207–216. [CrossRef] [PubMed]

MDPI

St. Alban-Anlage 66

4052 Basel

Switzerland

Tel. +41 61 683 77 34

Fax +41 61 302 89 18

www.mdpi.com

Agronomy Editorial Office

E-mail: agronomy@mdpi.com

www.mdpi.com/journal/agronomy

www.ingramcontent.com/pod-product-compliance
Lightning Source LLC
Chambersburg PA
CBHW051905210326
41597CB00033B/6026